Holt Mathematics

State Test Prep Workbook
Grade 6

HOLT, RINEHART AND WINSTON
A Harcourt Education Company

Orlando • Austin • New York • San Diego • London

Copyright © by Holt, Rinehart and Winston

All rights reserved. No part of this publication may be reproduced or transmitted in any form or by any means, electronic or mechanical, including photocopy, recording, or any information storage and retrieval system, without permission in writing from the publisher.

Teachers using Holt Mathematics may photocopy complete pages in sufficient quantities for classroom use only and not for resale.

HOLT and the **"Owl Design"** are trademarks licensed to Holt, Rinehart and Winston, registered in the United States of America and/or other jurisdictions.

Printed in the United States of America

If you have received these materials as examination copies free of charge, Holt, Rhinehart and Winston retains title to the materials and they may not be resold. Resale of examination copies is strictly prohibited and is illegal.

Possesion of this publication in print format does not entitle user to convert this publication, or any portion of it, into electronic format.

ISBN 0-03-078273-2

9 10 11 12   1421   12 11
4500294576

CONTENTS

Diagnostic Test 1
Diagnostic Test Answer Sheet 11
NAEP Objective Test 8.1.1.e 12
NAEP Objective Test 8.1.1.f 14
NAEP Objective Test 8.1.1.i 16
NAEP Objective Test 8.1.2.b 18
NAEP Objective Test 8.1.3.a 20
NAEP Objective Test 8.1.3.d 22
NAEP Objective Test 8.1.3.g 24
NAEP Objective Test 8.1.4.a 26
NAEP Objective Test 8.1.4.c 28
NAEP Objective Test 8.1.4.d 30
NAEP Objective Test 8.1.5.b 32
NAEP Objective Test 8.1.5.d 34
NAEP Objective Test 8.1.5.e 36
NAEP Objective Test 8.2.1.h 38
NAEP Objective Test 8.2.1.j 40
NAEP Objective Test 8.2.1.k 42
NAEP Objective Test 8.2.1.l 44
NAEP Objective Test 8.2.2.b 46
NAEP Objective Test 8.2.2.d 48
NAEP Objective Test 8.2.2.f 50
NAEP Objective Test 8.3.1.b 52
NAEP Objective Test 8.3.1.c 54
NAEP Objective Test 8.3.1.e 56
NAEP Objective Test 8.3.2.a 58
NAEP Objective Test 8.3.2.c 60
NAEP Objective Test 8.3.2.d 62
NAEP Objective Test 8.3.2.e 64
NAEP Objective Test 8.3.3.b 66
NAEP Objective Test 8.3.3.f 68
NAEP Objective Test 8.3.4.d 70

CONTENTS continued

NAEP Objective Test 8.4.1.a 72
NAEP Objective Test 8.4.1.c 74
NAEP Objective Test 8.4.1.d 76
NAEP Objective Test 8.4.2.a 78
NAEP Objective Test 8.4.2.d 80
NAEP Objective Test 8.4.4.b 82
NAEP Objective Test 8.4.4.e 84
NAEP Objective Test 8.4.4.g 86
NAEP Objective Test 8.4.4.j 88
NAEP Objective Test 8.5.1.a 90
NAEP Objective Test 8.5.1.c 92
NAEP Objective Test 8.5.1.e 94
NAEP Objective Test 8.5.2.c 96
NAEP Objective Test 8.5.2.d 98
NAEP Objective Test 8.5.2.g 100
NAEP Objective Test 8.5.3.b 102
NAEP Objective Test 8.5.3.c 104
NAEP Objective Test 8.5.4.a 106
NAEP Objective Test 8.5.4.c 108
NAEP Objective Test 8.5.4.e 110
Sample Test A 112
Sample Test B 116
Answer Sheet for Sample Tests 120

Diagnostic Test

Select the best answer for questions 1–60. Fill in the correct bubble on your answer sheet.

1. Marc bought an old car that needed many repairs. In the first 6 months he owned the car, he paid a mechanic $438, $602, $194, $311, and $275. Which is the best estimate of the total cost of the repairs?

   A $1,600
   B $1,700
   C $1,800
   D $1,900

2. Use the order of operations to simplify the expression below.
   $$75 - 3 \times (24 \div 6)^2 + 2$$
   F 29
   G 42
   H 50
   I 1,154

3. Which value is the missing number in the sequence?
   185 199 213 ____ 241 255
   A 226
   B 227
   C 228
   D 229

4. Nathan bought a rare coin for $13.00. The coin's value increased 75¢ each month after he bought the coin. What was the value of the coin after five months?

   F $16.00
   G $16.75
   H $17.50
   I $18.00

5. Henry is making a family tree. He has two parents, four grandparents, and eight great-grandparents. Following this pattern, how many great-great-grandparents will Henry have?

   A 8
   B 10
   C 12
   D 16

6. What decimal number represents the shaded portion of the grid below?

   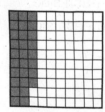

   F 0.0028
   H 0.028
   G 0.28
   I 2.8

Holt Mathematics Grade 6

Name _____ Date _____

Diagnostic Test continued

7. Lydia attends volleyball practice three times a week for two hours each session. Which expression describes the total number of hours Lydia attends volleyball practice each week?

   A $3 \cdot 2$
   B $3 \cdot 2 \cdot 7$
   C 3^2
   D $3 + 2$

8. What is the value of the expression $11r + 3s$ when $r = 5$ and $s = 7$?

   F 65
   G 76
   H 92
   I 152

9. Leah wants to solve the equation $2x - 7 = 65$. What step should she take first?

   A Divide each side by 2.
   B Subtract 7 from each side.
   C Add 7 to each side.
   D Add $2x$ to each side.

10. Rodeo riders are scored between 60 and 80 points for each ride. After three rides, the scores are totaled to see who wins. Which of these riders came in second?

    Points Earned

Ride	Cody	Dusty	Trampas	Bruce
1	71.5	74	70	68.5
2	73	68.5	73.5	74.5
3	72.5	70	72	73

    F Cody
    G Dusty
    H Trampas
    I Bruce

11. A jewelry designer has a supply of gold that has a mass of 1.438 kilograms. She has designed a gold bracelet with links that have a mass of 0.08 kilograms each. How many links can she make from her supply of gold?

    A 15
    B 15.6
    C 17
    D 17.975

12. It is estimated that the current population of the entire Earth is about 6,600,000,000. What is that number in scientific notation?

    F 6.6×10^9
    G 0.66×10^8
    H 6.6×10^8
    I 0.66×10^9

Diagnostic Test continued

13. Sandy's mother bought three super-giant pizzas and five large pizzas for Sandy's birthday party. The super-giant pizzas cost $19.00 each. The total cost for all the pizzas was $139.50. What was the cost of one large pizza?

    A $16.50   C $24.10
    B $17.00   D $82.50

14. In 1909, Alice Ramsey drove across the United States in her Maxwell car, the first woman to complete the trip. The trip took 60 days. What is the prime factorization of 60?

    F $2 \times 3 \times 5$ H $2^2 \times 3 \times 5$
    G $2^2 \times 3 \times 7$ I $2 \times 3^2 \times 5$

15. It is recommended that a person eat no more than 2,400 grams of sodium per day. Andy ate 864 grams of sodium for lunch. What percent of the recommended daily amount does this represent?

    A 33%   C 37%
    B 36%   D 41%

16. Mr. Hielo divided all of his 819 pencils into equal groups and put them into boxes. Which number of boxes could he NOT have used?

    F 3 boxes
    G 7 boxes
    H 4 boxes
    I 9 boxes

17. Which BEST explains why 19 is a prime number?

    A It is an odd number that has two digits.
    B It is a whole number less than 20.
    C It is not divisible by 2, 3, 5, 6 or 7.
    D It is a counting number that has exactly two factors, 1 and itself.

18. Flor wants to solve the equation $(\frac{3}{5})x = 16$. What step should she take first?

    F Subtract $\frac{3}{5}$ from each side.
    G Subtract 16 from each side.
    H Multiply both sides by 5.
    I Divide each side by 16.

19. In the graph below, if the value of x is increased by 1, what is the effect on the value of y?

    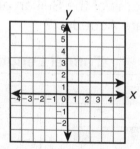

    A The value of y remains the same.
    B The value of y is increased by 1.
    C The value of y is increased by 2.
    D The value of y is increased by 5.

Diagnostic Test continued

20. The lengths of some boards in inches are: 58, 68, 70, 48, 60, 64, 62, 58. If another board 62 inches long were added to the list, what would happen to the median length?

    F It would stay the same.

    G It would increase by $\frac{1}{2}$.

    H It would increase by 1.

    I It would decrease by 1.

21. Joyce earned an average (mean) of $900 a week over four weeks. If she earned $875, $915, and $855 for the first three weeks, how much did she earn the fourth week?

    A $855 C $945
    B $895 D $955

22. The graph below shows the length in miles of some tunnels. Which is the best estimate of the difference in the lengths of the Seikan and the Delaware Aqueduct?

    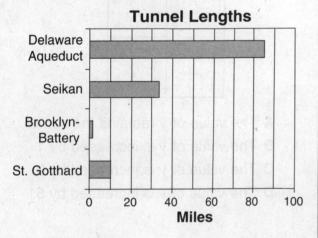
    Tunnel Lengths

    F 30 miles   H 60 miles
    G 50 miles   I 80 miles

23. Which would be the most appropriate way to display this data?

    Which is your dominant hand?

Right	62
Left	24
Ambidextrous	14

    A circle graph
    B box-and-whisker plot
    C line graph
    D scatter plot

24. Which statement is ALWAYS true for two similar triangles?

    F Their corresponding sides have equal lengths.

    G Their corresponding angles have equal measures.

    H Their corresponding sides have equal lengths and their corresponding angles have equal measures.

    I Only one of their corresponding angles have equal measures.

25. Raul goes on a three-day hiking trip with his father. Each day they hike for 6 hours at an average speed of 3 miles per hour. After three days, how many miles have Raul and his father hiked?

    A 9 miles
    B 18 miles
    C 27 miles
    D 54 miles

Diagnostic Test continued

26. An office building is 82 feet tall, 50 feet wide, and 140 feet long. What is the ratio of the length of the building to its height?
    F 41:70
    G 70:25
    H 70:41
    I 25:70

27. Paul put some money in a bank that paid 5% interest. Three years later, his account had $2,760. How much money did he originally put in the account?
    A $2,000
    B $2,250
    C $2,390
    D $2,400

28. The scale on a map is 2 inches = 25 miles. If two cities are 7 inches apart on a map, what is the actual distance between the cities?
    F 1.8 miles
    G 7.1 miles
    H 87.5 miles
    I 175 miles

29. Catherine bikes 3.5 miles to work each day in 25 minutes. If she bikes at the same speed, how long will it take her to bike 14 miles?
    A 100 minutes
    B 196 minutes
    C 4 hours
    D 10 hours

30. The circle graph shows the results of a survey asking voters whom they plan to vote for in an upcoming election. Which statement can be inferred from the poll?

    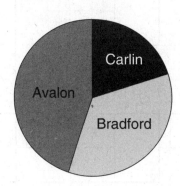

    F Carlin will not win.
    G Bradford will win.
    H No one will win.
    I Avalon cannot win.

31. There are 8 boys and 11 girls on the Harriso Middle School ice-skating team. Which fraction shows the ratio of boys to total team members?
    A $\frac{8}{11}$ C $\frac{11}{19}$
    B $\frac{11}{8}$ D $\frac{8}{19}$

Diagnostic Test continued

32. A lamppost is 8 meters tall and casts a shadow 20 meters long. At the same time, a sign beside the lamppost casts a shadow 6.5 meters long. What is the height of the sign?

    F 2.6 meters
    G 3.25 meters
    H 16.25 meters
    I 24.6 meters

33. Which pair of rectangles are similar?

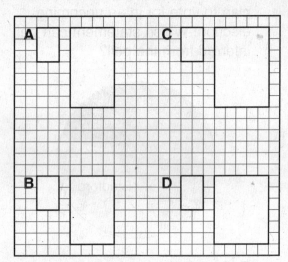

34. A company sells pins that are 1:6 scale models of Olympic medals. If the actual medal is 3.75 inches wide, how wide is the pin?

    F 0.00625 inches
    G 0.0625 inches
    H 0.625 inches
    I 0.675 inches

35. In which figure is the diagonal a line of symmetry?

    A   C

    B   D

36. Which figure has all congruent angles?

    F quadrilateral
    G scalene triangle
    H isosceles triangle
    I regular pentagon

37. When the point (−2, 2) is reflected across the y-axis, what are the new coordinates?

    A (−2, 2)
    B (2, −2)
    C (2, 2)
    D (−2, −2)

Diagnostic Test continued

38. Which property is true for all trapezoids?

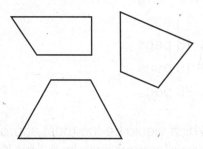

F One pair of sides is parallel.
G Adjacent sides are perpendicular.
H Two sides are congruent.
I Opposite angles are congruent.

39. What does this pentagon look like after it is rotated $\frac{1}{2}$ turn?

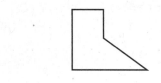

A 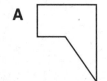   C

B 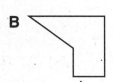   D

40. Which is true for all parallel lines?
F They intersect in one point.
G They intersect in two points.
H They intersect in different planes.
I They never intersect.

41. Julia is making a banner shaped like an isosceles triangle. Two sides measure 8 inches and 20 inches. What is the measure of the third side?
A 8 inches
B 20 inches
C 8 inches or 20 inches
D There is not enough information to answer the question.

42. If you cut a regular pentagon along one diagonal, what will you get?

F an isosceles triangle and a trapezoid
G an equilateral triangle and a trapezoid
H an isosceles triangle and a parallelogram
I an equilateral triangle and a parallelogram

Name _____ Date _____ Class _____

Diagnostic Test continued

43. If Sarah wants to measure the ribbon to wrap a present, which unit of measure should she use to get the most accurate measure?

   A kilometer
   B meter
   C centimeter
   D millimeter

44. The temperature at which water freezes is 32°F. At what temperature does water freeze in Celsius?

   F −32°C
   G 0°C
   H 32°C
   I 100°C

45. Which unit of measure is missing from this table?

Family Member	Height (in __?__)
Mr. Shella	1.75
Mrs. Shella	1.6
Amy	1.52
Brandon	1.15

   A centimeters
   B feet
   C inches
   D meters

46. How many 3-inch long pegs can Evan cut from a wooden rod 5 feet long?

   F 7 pegs
   G 15 pegs
   H 16 pegs
   I 20 pegs

47. Which would be the most appropriate measurement instrument for Kerry to use when she cuts lumber for a house she is building?

   A stop watch
   B balance scale
   C tape measure
   D protractor

48. Which number is frequently used to represent π?

   F 31.4
   G 3.14
   H 3.41
   I $2r$

Diagnostic Test continued

49. Which statement describes this solid?

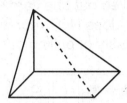

   A It has 4 edges.
   B It has 8 vertices.
   C It has 4 triangular faces.
   D It has a triangular base.

50. The base of a triangular sail is 4 feet long, and its height is 15 feet. What is the area of the sail?

   F 15 square feet
   G 30 square feet
   H 38 square feet
   I 60 square feet

51. Which solid does NOT have a triangular face or base?

   A 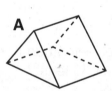   C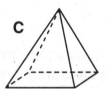

   B 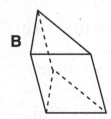   D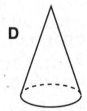

52. What is the volume of this swimming pool? Use 3.14 for π.

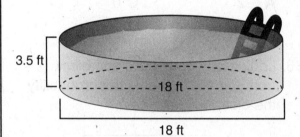

   F 197.82 cubic feet
   G 283.5 cubic feet
   H 692.37 cubic feet
   I 890.19 cubic feet

53. Reggie had $415 in his checking account. By mistake, he wrote a check for $482. What is the balance in his account now?

   A −$67
   B −$57
   C $57
   D $67

54. Which point is NOT the same distance from the origin as (−3, 4)?

   F (3, −4)
   G (−3, −4)
   H (3, 4)
   I (−3, 0)

Diagnostic Test continued

55. What is the y-coordinate of the ordered pair (1,−2)?
A −2
B −1
C 0
D 1

56. Louisa randomly pulls one block from the box. What is the probability the block will be black?

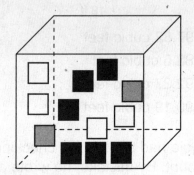

F $\frac{1}{3}$
G $\frac{1}{2}$
H $\frac{5}{6}$
I 1

57. A bag contains 5 orange, 4 green, and 6 red marbles. One marble is pulled at random from the bag. How many different outcomes are in the sample space for this experiment?
A 1
B 3
C 6
D 15

58. Ryan placed these cards in a bag. He pulled out one card at random. Which does NOT represent the probability of pulling out the letter I?

F 20%
G $\frac{1}{5}$
H 0.2
I 2%

59. There is a 75% chance of thunderstorms today. What is the probability that there will NOT be any thunderstorms today?
A 25%
B 50%
C 75%
D 100%

60. Which shows the following numbers in order from greatest to least?

0.001, −0.01, −0.11, −0.101, 1.01, −1.11

F −1.11, −0.11, −0.101, −0.01, 0.001, 1.01
G −1.11, 1.01, −0.11, −0.101, −0.01, 0.001
H 1.01, 0.001, −0.01, −0.101, −0.11, −1.11
I 1.01, 0.001, −1.11, −0.11, −0.101, −0.01

Name _____ Date _____ Class _____

Diagnostic Test Answer Sheet

1. Ⓐ Ⓑ Ⓒ Ⓓ
2. Ⓕ Ⓖ Ⓗ Ⓘ
3. Ⓐ Ⓑ Ⓒ Ⓓ
4. Ⓕ Ⓖ Ⓗ Ⓘ
5. Ⓐ Ⓑ Ⓒ Ⓓ
6. Ⓕ Ⓖ Ⓗ Ⓘ
7. Ⓐ Ⓑ Ⓒ Ⓓ
8. Ⓕ Ⓖ Ⓗ Ⓘ
9. Ⓐ Ⓑ Ⓒ Ⓓ
10. Ⓕ Ⓖ Ⓗ Ⓘ
11. Ⓐ Ⓑ Ⓒ Ⓓ
12. Ⓕ Ⓖ Ⓗ Ⓘ
13. Ⓐ Ⓑ Ⓒ Ⓓ
14. Ⓕ Ⓖ Ⓗ Ⓘ
15. Ⓐ Ⓑ Ⓒ Ⓓ
16. Ⓕ Ⓖ Ⓗ Ⓘ
17. Ⓐ Ⓑ Ⓒ Ⓓ
18. Ⓕ Ⓖ Ⓗ Ⓘ
19. Ⓐ Ⓑ Ⓒ Ⓓ
20. Ⓕ Ⓖ Ⓗ Ⓘ
21. Ⓐ Ⓑ Ⓒ Ⓓ
22. Ⓕ Ⓖ Ⓗ Ⓘ
23. Ⓐ Ⓑ Ⓒ Ⓓ
24. Ⓕ Ⓖ Ⓗ Ⓘ
25. Ⓐ Ⓑ Ⓒ Ⓓ
26. Ⓕ Ⓖ Ⓗ Ⓘ
27. Ⓐ Ⓑ Ⓒ Ⓓ
28. Ⓕ Ⓖ Ⓗ Ⓘ
29. Ⓐ Ⓑ Ⓒ Ⓓ
30. Ⓕ Ⓖ Ⓗ Ⓘ

31. Ⓐ Ⓑ Ⓒ Ⓓ
32. Ⓕ Ⓖ Ⓗ Ⓘ
33. Ⓐ Ⓑ Ⓒ Ⓓ
34. Ⓕ Ⓖ Ⓗ Ⓘ
35. Ⓐ Ⓑ Ⓒ Ⓓ
36. Ⓕ Ⓖ Ⓗ Ⓘ
37. Ⓐ Ⓑ Ⓒ Ⓓ
38. Ⓕ Ⓖ Ⓗ Ⓘ
39. Ⓐ Ⓑ Ⓒ Ⓓ
40. Ⓕ Ⓖ Ⓗ Ⓘ
41. Ⓐ Ⓑ Ⓒ Ⓓ
42. Ⓕ Ⓖ Ⓗ Ⓘ
43. Ⓐ Ⓑ Ⓒ Ⓓ
44. Ⓕ Ⓖ Ⓗ Ⓘ
45. Ⓐ Ⓑ Ⓒ Ⓓ
46. Ⓕ Ⓖ Ⓗ Ⓘ
47. Ⓐ Ⓑ Ⓒ Ⓓ
48. Ⓕ Ⓖ Ⓗ Ⓘ
49. Ⓐ Ⓑ Ⓒ Ⓓ
50. Ⓕ Ⓖ Ⓗ Ⓘ
51. Ⓐ Ⓑ Ⓒ Ⓓ
52. Ⓕ Ⓖ Ⓗ Ⓘ
53. Ⓐ Ⓑ Ⓒ Ⓓ
54. Ⓕ Ⓖ Ⓗ Ⓘ
55. Ⓐ Ⓑ Ⓒ Ⓓ
56. Ⓕ Ⓖ Ⓗ Ⓘ
57. Ⓐ Ⓑ Ⓒ Ⓓ
58. Ⓕ Ⓖ Ⓗ Ⓘ
59. Ⓐ Ⓑ Ⓒ Ⓓ
60. Ⓕ Ⓖ Ⓗ Ⓘ

Copyright © by Holt, Rinehart and Winston.
All rights reserved.

Holt Mathematics Grade 6

NAEP Objective Test 8.1.1.e

Select the best answer for questions 1–12. Fill in the correct bubble on your answer sheet.

1. Sandy lost $1\frac{3}{8}$ pounds training for basketball season. What is that number in decimal form?

   A 0.375 pounds
   B 1.3 pounds
   C 1.375 pounds
   D 1.40 pounds

2. Which is NOT another way to write the part of a kilogram that this container of yogurt holds?

   F $\frac{17}{10}$
   G 0.170
   H $\frac{1}{10} + \frac{7}{100}$
   I $\frac{17}{100}$

3. Agnes got 43 of 50 questions on the math test correct. What is her score expressed as a percent?

   A 7%
   B 43%
   C 50%
   D 86%

4. The label on a gallon of whole milk says that the milk contains 3.3% fat. What is that number expressed as a decimal?

   F 0.0033
   G 0.033
   H 0.33
   I 3.3

5. The trip meter on Mr. Nesmith's car shows the number of miles he drove to work.

   Which shows another way to describe this distance?

   A $12\frac{2}{5}$ miles
   B $12\frac{4}{5}$ miles
   C $12\frac{1}{8}$ miles
   D $12\frac{2}{25}$ miles

6. Which of these lists does NOT show a set of equivalent numbers?

   F 0.4, $\frac{10}{25}$, 0.40
   G $\frac{1}{4}$, $\frac{5}{20}$, 0.25
   H 1.64, $1\frac{32}{50}$, $1\frac{16}{25}$
   I $5\frac{2}{50}$, 5.06, $5\frac{6}{100}$

NAEP Objective Test 8.1.1.e continued

7. The scale shows the weight of Beth's dog in pounds.

Which is another way to describe this weight?

A $24\frac{3}{5}$ pounds

B $24\frac{1}{6}$ pounds

C $24\frac{3}{8}$ pounds

D $24\frac{3}{50}$ pounds

8. Three-fifths of Jake's football team is on the honor roll. What percent of the team in on the honor roll?

F 3.5%

G 35%

H 60%

I 70%

9. During the championship baseball game, DeWayne hit the ball 12 times. Nine of those hits were foul balls. What percent of his hits were foul balls?

A 3%

B 12%

C 75%

D 120%

10. Which of these shows a set of equivalent numbers?

F 3.09, $3\frac{9}{10}$, 3.9%

G $2\frac{42}{50}$, 2.84, $2\frac{21}{25}$

H 8%, 0.8, $\frac{20}{25}$

I $\frac{16}{25}$, 32% 0.64

11. The weather forecaster said that the chance of rain today was 40%. Which shows another way to describe this number?

A 0.04

B $\frac{40}{10}$

C 4.0

D $\frac{10}{25}$

12. The CD Sophie uses to record her homework holds 600 MB of data. Her homework uses 120 MB of that space. What percent of the CD is used?

F 12%

G 20%

H 60%

I 85%

Name _____ Date _____ Class _____

NAEP Objective Test 8.1.1.f

Select the best answer for questions 1–12. Fill in the correct bubble on your answer sheet.

1. The Mississippi River runs 2,340 miles. Which of the following shows the number of miles written in scientific notation?
   A 23.4×10^2
   B 2.34×10^2
   C 2.34×10^3
   D 0.234×10^4

2. The population of the United States in July, 2005, was estimated to be about 295,000,000. Which of the following shows that number in scientific notation?
   F 2.95×10^8
   G 2.95×10^9
   H 29.5×10^7
   I 29.5×10^8

3. The distance from the planet Jupiter to the Sun averages 4.8×10^8 miles. What is the number of miles in standard notation?
   A 480,000
   B 4,800,000
   C 48,000,000
   D 480,000,000

4. The state of Alaska incudes about 129,000,000 acres of forested land. Which of the following shows the number of acres in scientific notation?
   F 1.29×10^7
   G 1.29×10^8
   H 12.9×10^8
   I 12.9×10^9

5. Mount McKinley is the highest point in North America at about 20,000 feet. Which of the following shows the number of feet in scientific notation?
   A 1.0×20^4
   B 2.0×10^3
   C 2.0×10^4
   D 2.0×10^5

6. Victoria Island in Canada has an area of about 8.4×10^4 square miles. Which of the following shows the number of square miles in standard notation?
   F 84
   G 840
   H 8,400
   I 84,000

NAEP Objective Test 8.1.1.f continued

7. The approximate length of the Earth's equator is shown in the diagram.

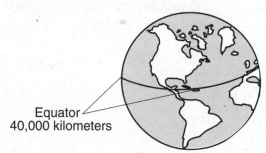

Equator
40,000 kilometers

Which of the following shows the number of kilometers in scientific notation?

A 4.0×10^3
B 4.0×10^4
C 4.0×10^5
D 4.0×10^6

8. The area of the Atlantic Ocean is about 33 million square miles. Which of the following shows the number of square miles in scientific notation?

F 0.33×10^7
G 0.33×10^8
H 3.3×10^6
I 3.3×10^7

9. One International Nautical Mile is equal to just over 6.076×10^3 feet. Which of the following shows the number of feet in standard form?

A 607.6
B 6,076
C 60,760
D 607,600

10. The population of Tokyo, Japan is about 12,400,000. Which of the following shows that number in scientific notation?

F 0.124×10^9
G 1.24×10^6
H 1.24×10^7
I 12.4×10^7

11. The world record for the land speed record was set by a jet car traveling at 7.63×10^2 miles per hour. Which of the following shows the number of miles per hour written in standard notation?

A 763
B 7,630
C 76,300
D 763,000

12. The census of 2000 found that more than 6.42×10^5 people lived in North Dakota. Which of the following shows that number in standard form?

F 6,420
G 64,200
H 642,000
I 6,420,000

Name _____ Date _____ Class _____

NAEP Objective Test 8.1.1.i

Select the best answer for questions 1–12. Fill in the correct bubble on your answer sheet.

1. The lengths in miles of some major rivers are shown in the chart.

Heilong River	Irtish River
2,758 miles	2,704 miles
Parana River	Zaire River
2,795 miles	2,716 miles

    Which of the following shows those lengths in order from least to greatest?

    A 2,704; 2,716; 2,795; 2,758
    B 2,795; 2,758; 2,716; 2,704
    C 2,704; 2,758; 2,716; 2,795
    D 2,704; 2,716; 2,758; 2,795

2. Jacob collected some rocks for a rock garden. The rocks had masses of 13.05, 13.55, 13.055, and 13.5 kilograms. Which shows the masses of the rocks in order from greatest to least?

    F 13.55; 13.5; 13.055; 13.05
    G 13.55; 13.055; 13.5; 13.05
    H 13.05; 13.055; 13.5; 13.55
    I 13.05; 13.5; 13.055; 13.55

3. Which of the following shows the mixed numbers in order from least to greatest?

    A $2\frac{3}{4}, 2\frac{5}{8}, 2\frac{4}{7}, 2\frac{7}{10}$
    B $2\frac{4}{7}, 2\frac{5}{8}, 2\frac{7}{10}, 2\frac{3}{4}$
    C $2\frac{3}{4}, 2\frac{4}{7}, 2\frac{5}{8}, 2\frac{7}{10}$
    D $2\frac{4}{7}, 2\frac{7}{10}, 2\frac{3}{4}, 2\frac{5}{8}$

4. Four friends played a video game. Troy scored 24,538 points, Lauren scored 24,358 points, Matt scored 25,338 points, and Becky scored 23,485 points. Who won the game?

    F Becky
    G Lauren
    H Matt
    I Troy

5. In the game of golf, the lowest score wins. The chart shows the average score of four golfers for the summer.

Tom	Bill	Fred	Cory
69.24	68.42	68.04	68.02

    Who was the best golfer that summer?

    A Bill
    B Cory
    C Fred
    D Tom

6. Which of the following shows the decimals in order from greatest to least?

    F 0.07; 0.67; 0.067; 0.607
    G 0.067; 0.67; 0.07; 0.607
    H 0.607; 0.67; 0.07; 0.067
    I 0.67; 0.607; 0.07; 0.067

16 Holt Mathematics Grade 6

NAEP Objective Test 8.1.1.i continued

7. The team of scientists in the Arctic recorded the following temperatures at their base camp.

Day 1	Day 2	Day 3	Day 4
−34°	−24°	−29°	−36°

Which day was the warmest day?
A Day 1
B Day 2
C Day 3
D Day 4

8. Marco collects stamps. He has stamps that are $\frac{3}{4}$ inches wide, $\frac{3}{5}$ inches wide, $\frac{5}{8}$ inches wide, and $\frac{4}{7}$ inches wide. Which shows the stamps in order from least to greatest width?

F $\frac{4}{7}, \frac{3}{5}, \frac{5}{8}, \frac{3}{4}$

G $\frac{3}{4}, \frac{3}{5}, \frac{4}{7}, \frac{5}{8}$

H $\frac{3}{4}, \frac{3}{5}, \frac{5}{8}, \frac{4}{7}$

I $\frac{4}{7}, \frac{5}{8}, \frac{3}{5}, \frac{3}{4}$

9. When ordering the fractions $\frac{1}{3}, \frac{4}{5}, \frac{7}{8},$ and $\frac{1}{6}$, which common denominator should be used?
A 40
B 80
C 120
D 720

10. During the gymnastics meet, Danielle received the following scores for her performance on the balance beam.

Judge A	Judge B	Judge C	Judge D
8.97	8.87	8.78	8.98

Which judge gave her the least score?
F Judge A
G Judge B
H Judge C
I Judge D

11. Mars is at least 205,000,000 miles from the Sun, Neptune is at least 2,770,000,000 miles from the Sun, Saturn is at least 840,000,000 miles from the Sun, and Jupiter is at least 460,000,000 miles from the Sun. Which planet is the greatest distance from the Sun?
A Mars
B Neptune
C Saturn
D Jupiter

12. Which of the following shows the integers in order from greatest to least?
F −15, 8, −4, 2
G 2, −4, 8, −15
H −15, −4, 2, 8
I 8, 2, −4, −15

Holt Mathematics Grade 6

NAEP Objective Test 8.1.2.b

Select the best answer for questions 1–12. Fill in the correct bubble on your answer sheet.

1. The table shows the number of people who entered the county fair through each gate on Saturday. About how many people in all attended the fair on Saturday?

   Saturday's Fair Attendance

Gate	Number of People
A	9,274
B	4,830
C	3,149
D	6,502

   A 20,000
   B 22,000
   C 24,000
   D 26,000

2. This week Mr. Mabon used 38 gallons of gas to drive 826 miles on business. Which is the best estimate of the fuel efficiency of his car?

   F 20 miles per gallon
   G 30 miles per gallon
   H 40 miles per gallon
   I 50 miles per gallon

3. Which is the best estimate of $\frac{2}{3} + \frac{4}{5}$?

   A 0
   B 1
   C 2
   D 3

4. Khori jogs a distance of 0.66 miles around his block. His goal is to jog 14 miles per week. About how many trips around the block will he need to make to reach his goal?

   F 15
   G 21
   H 32
   I 38

5. A local bank has $30,000 at the beginning of the day. During the day $19,761 is withdrawn and $31,012 is deposited. Estimate how much money the bank has at the end of the day.

   A $20,000
   B $30,000
   C $40,000
   D $50,000

6. Laurel Valley School has 27 buses. An average of 52 students ride each bus. About how many students at Laurel Valley School ride the bus?

   F 1,000
   G 1,500
   H 1,800
   I 15,000

NAEP Objective Test 8.1.2.b continued

7. Mrs. Benson ordered 7 large pizzas for her son's birthday party. After the party, only $1\frac{2}{3}$ of a pizza remained.

   About how much pizza was eaten at the party?

   A 4 pizzas
   B 5 pizzas
   C 6 pizzas
   D 7 pizzas

8. A spool of chain originally contained 50 yards of chain. The hardware store sold two pieces of the chain, one 3.6 yards long and one 12.05 yards long. About how much chain is left on the spool?

   F 16 yards
   G 32 yards
   H 34 yards
   I 38 yards

9. Which is the best estimate of $4\frac{7}{9} - 2\frac{1}{8}$?

   A 1
   B 2
   C 3
   D 4

10. Mrs. Jillian's car gets 31.4 miles to the gallon of gas. If Mrs. Jillian puts 15.3 gallons of gas in her car, about how far can she expect to drive?

    F 450 miles
    G 500 miles
    H 550 miles
    I 600 miles

11. Which is the best estimate of 698.70 + 148.09?

    A 700
    B 750
    C 800
    D 850

12. The Corner Market sells grapes in bunches. The bunches have the following weights.

    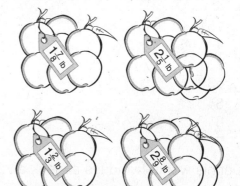

    About how many pounds of grapes are there in all?

    F 7 pounds
    G 8 pounds
    H 9 pounds
    I 10 pounds

NAEP Objective Test 8.1.3.a

Select the best answer for questions 1–12. Fill in the correct bubble on your answer sheet.

1. Chantal is thinking of buying one of these backpacks for school. How much more is the one with wheels than the one without wheels?

   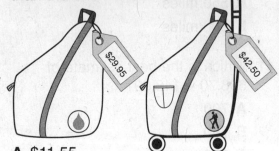

   A $11.55
   B $12.55
   C $12.65
   D $72.45

2. What is $\frac{3}{4}$ of $\frac{16}{21}$?

   F $\frac{8}{15}$
   G $\frac{4}{7}$
   H $\frac{63}{64}$
   I $\frac{8}{7}$

3. Dennis works two days a week as a cashier. He works 7 hours on Saturday and $5\frac{1}{4}$ hours on Monday. How many more hours does Dennis work on Saturday than on Monday?

   A $1\frac{3}{4}$ hours
   B $1\frac{9}{4}$ hours
   C $2\frac{1}{4}$ hours
   D $12\frac{1}{4}$ hours

4. The Wong family bought 2 adult tickets and 4 student tickets for a concert. The total cost of the tickets without tax was $120. What is the price of one student ticket?

   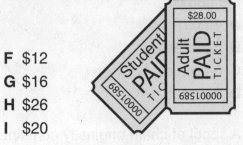

   F $12
   G $16
   H $26
   I $20

5. Jeff wants to save money for college. He plans to save 25% of his earnings. How much will he save if he earns $216?

   A $25
   B $50
   C $52
   D $54

6. Divers discovered a shipwreck lying 54 feet below sea level. The crane they rented to hoist up pieces of the wreck stood 87 feet above sea level. What is the distance between the level of the wreck and the level of the crane?

   F 33 feet
   G 131 feet
   H 141 feet
   I 151 feet

NAEP Objective Test 8.1.3.a continued

7. The sixth grade classes at one school collected cans for a food drive. The chart shows the number of pounds of cans each class collected.

Room 203	Room 210	Room 214	Room 218
$56\frac{5}{16}$ pounds	$47\frac{7}{12}$ pounds	$59\frac{1}{4}$ pounds	$52\frac{3}{8}$ pounds

How many pounds were collected in all?

A $215\frac{25}{48}$ pounds

B $216\frac{25}{48}$ pounds

C $216\frac{9}{16}$ pounds

D $216\frac{31}{48}$ pounds

8. Eloise wants to buy a DVD player that costs $136.95. She plans to save $27.39 each month. At this rate, how many months will it take Eloise to save enough money?

F 5 months
G 6 months
H 7 months
I 8 months

9. What is 15% of 107?

A 1.605
B 16.05
C 160.5
D 1,605

10. Jack got a score of 92% on his math test. If there were 25 questions on the test, how many did he get correct?

F 22
G 23
H 24
I 25

11. What is the sum of -234 and 435?

A -669
B -192
C 201
D 669

12. On a map, two cities are shown as being 3.4 inches apart. The map scale says that 1 inch is equal to 7.5 miles. What is the actual distance between these cities?

F 10.9 miles
G 11.3 miles
H 22 miles
I 25.5 miles

Name _____ Date _____ Class _____

NAEP Objective Test 8.1.3.d

Select the best answer for questions 1–12. Fill in the correct bubble on your answer sheet.

1. Sarah wants to make $2\frac{1}{2}$ recipes of Play Clay for her daughter's birthday party. How much salt will she need in all?

   Play Clay Recipe

Flour	Salt	Soda	Water
$2\frac{1}{4}$ cup	$3\frac{2}{3}$ tspns	$1\frac{1}{3}$ cup	$\frac{7}{8}$ cup

   A $2\frac{2}{3}$ teaspoons
   B $5\frac{5}{6}$ teaspoons
   C $8\frac{1}{2}$ teaspoons
   D $9\frac{1}{6}$ teaspoons

2. A jar of peanuts says it contains 12 ounces of peanuts. Emily ate 0.4 of the peanuts. How many ounces of peanuts are left?

   F 0.48 ounces
   G 0.72 ounces
   H 4.8 ounces
   I 7.2 ounces

3. Which is the product of -35×562?

   A $-19{,}670$
   B $-18{,}450$
   C $18{,}450$
   D $19{,}670$

4. Bob wants to cut a 32-inch board into pieces $6\frac{4}{5}$ inches long. How many small pieces can he make?

   F 3
   G 4
   H 5
   I 6

5. Directions for Neat-N-Clean soap call for $\frac{2}{5}$ pint of soap for one gallon of water. Would more or less than 1 pint be needed for 7 gallons of water? How much would be needed?

   A less, $\frac{17}{35}$ pint
   B more, $1\frac{4}{5}$ pint
   C more, $2\frac{4}{5}$ pint
   D more, $3\frac{1}{5}$ pint

6. What is the quotient of $-336 \div 56$? Is it more or less than 56?

   F -6, less
   G -5, more
   H 5, more
   I 6, less

Copyright © by Holt, Rinehart and Winston.
All rights reserved.

Holt Mathematics Grade 6

NAEP Objective Test 8.1.3.d continued

7. What is the difference between the product of 0.03 × 427 and the product of 0.3 × 427?
   A 11.529
   B 100
   C 115.29
   D 1,000

8. Mrs. Daniels made $3\frac{1}{5}$ kilograms of mashed potatoes for a big family dinner. Mr. Daniels ate $\frac{1}{8}$ of the potatoes. How many kilograms of potatoes did Mr. Daniels eat?
   F $\frac{3}{8}$ kilogram
   G $\frac{2}{5}$ kilogram
   H $\frac{17}{40}$ kilogram
   I $\frac{9}{20}$ kilogram

9. Find the product: −46 × −153.
   A −7,058
   B −7,038
   C 7,038
   D 7,058

10. Alicia got all the questions on the math test correct except one. She incorrectly found the quotient of −238 ÷ −14 to be 15. What is the correct quotient?
    F −15
    G −14
    H 14
    I 17

11. Derek sold 53 boxes of note cards to raise money for the chess team. Each box weighed 1.45 pounds. How many pounds of note cards will Derek have to deliver?
    A 1.98 pounds
    B 54.45 pounds
    C 74.25 pounds
    D 76.85 pounds

12. How many books that are $\frac{5}{6}$ inches wide will fit on a shelf that is 15 inches long?
    F 18
    G $18\frac{1}{3}$
    H 19
    I $19\frac{1}{6}$

NAEP Objective Test 8.1.3.g

Select the best answer for questions 1–12. Fill in the correct bubble on your answer sheet.

1. According to the chart, how much material would Jamie need to make a skirt and a matching cape?

Material Needed

Skirt	Shirt	Shorts	Cape
$2\frac{5}{6}$ yards	$1\frac{7}{8}$ yards	$1\frac{2}{3}$ yards	$4\frac{1}{6}$ yards

   A $6\frac{1}{2}$ yards
   B 7 yards
   C $7\frac{1}{2}$ yards
   D 8 yards

2. Each student in the sixth grade brought in $3.75 for a field trip to the museum. If there are 211 students in the sixth grade, about how much money was brought in for the trip?

   F $600
   G $700
   H $800
   I $1,100

3. It took Stephen $3\frac{1}{2}$ hours to do the yardwork. He spent $\frac{2}{3}$ of that time mowing the grass. How long did it take Stephen to mow the grass?

   A $1\frac{2}{3}$ hours
   B 2 hours
   C $2\frac{1}{3}$ hours
   D 3 hours

4. Sam bought six movie tickets for $34.50. How much did one ticket cost?

   F $5.25
   G $5.50
   H $5.75
   I $6.00

5. Gregg spends 105 minutes exercising every day. He spends $\frac{3}{5}$ of the time running. How many minutes does Gregg spend running each day?

   A 57 minutes
   B 60 minutes
   C 63 minutes
   D 66 minutes

6. A submarine is 450 meters below sea level. An airplane is 832 meters above sea level. What is the total distance between the plane and the submarine?

   F −382 meters
   G 382 meters
   H 641 meters
   I 1,282 meters

NAEP Objective Test 8.1.3.g continued

7. What is the difference between the temperature at 7 AM and the temperature at 3 PM?

Time	7 AM	11 AM	3 PM
Temperature	−4°F	7°F	19°F

  A 17°F
  B 21°F
  C 23°F
  D 25°F

8. Sarah's book bag weighs $18\frac{1}{3}$ pounds. When she removes her math textbook, the book bag weighs $16\frac{3}{8}$ pounds. How much does Sarah's math textbook weigh?

  F $1\frac{23}{24}$ pounds
  G $2\frac{1}{24}$ pounds
  H $2\frac{3}{8}$ pounds
  I $2\frac{23}{24}$ pounds

9. Sharonna bought a DVD for $18.98, a sweater for $24.97, a bracelet for $19.38, and a bottle of shampoo for $3.79. About how much did she spend in all?

  A $60.00
  B $65.00
  C $67.00
  D $69.00

10. When Nick entered the fifth grade, he was $58\frac{1}{2}$ inches tall. When he entered sixth grade, Nick had grown $4\frac{3}{4}$ inches. How tall was Nick when he entered sixth grade?

  F $62\frac{1}{4}$ inches
  G $62\frac{1}{2}$ inches
  H 63 inches
  I $63\frac{1}{4}$ inches

11. The art teacher found partially used boxes of clay that weighed $2\frac{3}{4}$ pounds, $\frac{1}{3}$ pound, $1\frac{5}{6}$ pounds, and $\frac{7}{12}$ pound. About how much clay did she find in all?

  A 3 pounds
  B 6 pounds
  C 7 pounds
  D 4 pounds

12. Mr. Williams lost $175 each month on the stock market over the last 9 months. What was the overall change in the value of his stock?

  F −$19.44
  G −$184.00
  H −$1,557.00
  I −$1,575.00

NAEP Objective Test 8.1.4.a

Select the best answer for questions 1–12. Fill in the correct bubble on your answer sheet.

1. Jason washed 6 mini-vans, 9 cars, 3 SUVs, and 1 motorcycle to raise money to buy his mother a present. What is the ratio of vans washed to total number of vehicles washed?

   A 6:9
   B 19:6
   C 13:6
   D 6:19

2. At a restaurant, the cook mixes 3 cups of iced tea mix with 2 cups of lemon juice and 5 gallons of water. What is the ratio of lemon juice to iced tea mix?

   F 3:4
   G 2:3
   H 4:5
   I 4:80

3. Of the 32 sixth-graders, 14 have brown hair. What is the ratio of students with brown hair to students who do not have brown hair?

   A 7:9
   B 14:32
   C 16:9
   D 16:7

4. In a can of mixed nuts, Ellen discovered there were 56 peanuts, 45 cashews, 44 almonds, and 19 pistachios. What is the ratio of almonds to all the nuts?

   F 44:56
   G 45:44
   H 11:41
   I 44:162

5. Granny's Groceries keeps track of the number of boxes of cereal they sell.

Yum-Ees	Oh, Nuts!	Bland-Os
132	354	103

   What is the ratio of Oh, Nuts! to the other cereals sold?

   A 354:132
   B 354:235
   C 235:354
   D 103:354

6. What is the ratio of shaded squares to the number of unshaded squares?

   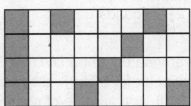

   F 32:10
   G 22:10
   H 16:16
   I 5:11

NAEP Objective Test 8.1.4.a continued

7. A football team has 11 offensive positions, 11 defensive positions, and 6 special-teams positions. What is the ratio of special-teams positions to all the positions?

   A 6:28
   B 6:11
   C 28:6
   D 11:28

8. The zoo has an exhibit with 32 penguins, 12 walruses, and 24 otters. What is the ratio of penguins to walruses and otters?

   F 9:8
   G 32:24
   H 8:9
   I 5:8

9. What is the ratio of the shaded shapes to the unshaded shapes?

   A 8:17
   B 9:17
   C 8:9
   D 9:8

10. A board game has a total of 120 tiles. There are an equal number of green, red, yellow, and blue tiles. What is the ratio of blue tiles to red and yellow tiles?

    F 80:40
    G 40:60
    H 2:3
    I 1:2

11. A can of tomato soup contains 20 grams of carbohydrates, 0 grams of fat, 2 grams of protein, and 12 grams of sugar. What is the ratio of grams of protein to grams of fat and sugar?

    A 12:2
    B 1:6
    C 2:20
    D 1:10

12. The Community Singers Group is made up of 13 adult men, 16 adult women, 12 boys, and 17 girls. What is the ratio of males to females?

    F 5:6
    G 25:33
    H 5:8
    I 33:25

NAEP Objective Test 8.1.4.c

Select the best answer for questions 1–12. Fill in the correct bubble on your answer sheet.

1. A chain saw uses a mixture of gas and oil. The correct proportion of gas to oil is show in the chart. Which number of pints of oil correctly completes the chart?

Gallons of Gas	6	15	27
Pints of Oil	2	5	?

   A 7
   B 8
   C 9
   D 10

2. The scale on a map states that "1 inch = 15 miles." How many inches are shown on the map for two towns that are 195 miles apart?

   F 13 inches
   G 14 inches
   H 15 inches
   I 17 inches

3. Solve the proportion $\frac{42}{7} = \frac{126}{x}$.

   A $x = 20.4$
   B $x = 21$
   C $x = 21.4$
   D $x = 22$

4. An architect's drawing of a house shows a doorway that is 3.5 inches tall. If the scale of the drawing is 1 inch = 2 feet, how tall will the door be on the actual house?

   F 6 feet
   G 6 feet, 8 inches
   H 6 feet, 10 inches
   I 7 feet

5. Amanda paid $4.95 for 2.5 pounds of sausage. What is the cost of 8 pounds of sausage?

   A $1.98
   B $9.90
   C $15.84
   D $16.40

6. Mr. Santos drove 345 miles on 15 gallons of gas. At that rate, how far could he expect to drive on 8 gallons of gas?

   F 23 miles
   G 184 miles
   H 207 miles
   I 234.8 miles

NAEP Objective Test 8.1.4.c continued

7. The scale of a model car is listed on the box as "$\frac{1}{18}$." The wheel on the car is 1.08 inches tall. How tall is the wheel on the actual car?
   A 19.44 inches
   B 19.54 inches
   C 19.84 inches
   D 194.4 inches

8. Solve the proportion $\frac{15}{x} = \frac{5}{9}$.
   F $x = 25$
   G $x = 26$
   H $x = 27$
   I $x = 28$

9. How many black dots should be added to the right so the ratios of white dots to black dots form a proportion?

   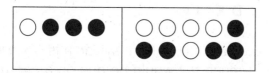

   A 5
   B 10
   C 15
   D 20

10. For their spaghetti supper, the Ladies Club figures that 4 boxes of spaghetti will serve 38 people. How many boxes of spaghetti will be needed to serve 836 people?
    F 80 boxes
    G 83 boxes
    H 85 boxes
    I 88 boxes

11. Emily painted 240 square feet of a room using $1\frac{1}{2}$ gallons of paint. At that rate, how many gallons of paint will it take to paint all 640 square feet of the room?
    A $2\frac{2}{3}$ gallons
    B $3\frac{1}{2}$ gallons
    C 4 gallons
    D $4\frac{3}{5}$ gallons

12. A dollmaker claims their Sassy Susie doll is a 3:8 scale duplicate of an average ten year-old girl. If the doll is 18.6 inches tall, how tall is the average ten year-old girl?
    F 49.6 inches
    G 49.8 inches
    H 74.4 inches
    I 148.8 inches

Name _____ Date _____ Class _____

NAEP Objective Test 8.1.4.d

Select the best answer for questions 1–12. Fill in the correct bubble on your answer sheet.

1. Davenport's Department Store sold a winter coat with a 35% discount. The original price of the coat was $89.99. Which was the sale price of the coat?

   A $31.49
   B $31.50
   C $58.49
   D $59.50

2. At a nice restaurant, Jack's father wanted to leave the waiter a 20% tip for his good service. The bill for the meal was $65.80. Which is the amount he should leave?

   F $11.74
   G $11.93
   H $12.00
   I $13.16

3. Mike had 568 stamps in his collection. He received 100 stamps for his birthday, and collected 42 more from friends. What percent increase does this represent?

   A 25%
   B 26%
   C 27.5%
   D 31%

4. The chart shows how much a store reduced the price of a sound system each week it was not sold. By what percent did the store reduce the price each week?

Week 1	Week 2	Week 3	Week 4
$468	$351	$263.25	$197.44

   F 20%
   G 22%
   H 25%
   I 30%

5. Joe put $4,000 in a savings account that paid 3.5% simple interest. At the end of one year, how much money was in the account?

   A $140
   B $4,040
   C $4,100
   D $4,140

6. Sales tax in Pennsylvania is 6%. How much tax would be due on a car that sells for $23,599?

   F $141.60
   G $1,415.94
   H $1,435.86
   I $1,522.91

NAEP Objective Test 8.1.4.d continued

7. Which is 59% of 5,287?
   A 311.933
   B 324.93
   C 3,117.52
   D 3,119.33

8. A package of dried beans says that 32% of the beans are pinto beans. If the package contains 462 beans, how many beans could be expected to be pinto beans?
   F 148
   G 149
   H 150
   I 151

9. A table saw is being sold with a discount of 20%. Mr. Lawry paid $432 for the saw after the discount. What was the original price of the saw?
   A $86.40
   B $345.60
   C $540.00
   D $600.00

10. A farm has 4 chickens, 3 horses, 5 goats, 6 cows, and 2 geese. What percent of the animals have only 2 legs?
    F 20%
    G 30%
    H 40%
    I 50%

11. Dennis placed $5,000 in a savings account last year. The account now holds $5,225. Which is the percent interest paid for this account?
    A 4%
    B 4.5%
    C 5%
    D 5.5%

12. Colin bought a suit that was on sale for $106.25. The original price for the suit was $125. Which percent discount was offered on the suit?
    F 15%
    G 18%
    H 20%
    I 25%

NAEP Objective Test 8.1.5.b

Select the best answer for questions 1–12. Fill in the correct bubble on your answer sheet.

1. Which is NOT a factor of 20?
   A 1
   B 5
   C 10
   D 40

2. What is the prime factorization of 245?
   F 7^3
   G $7^2 \times 5$
   H $3 \times 2^3 \times 5$
   I 7×35

3. What is the greatest common factor (GCF) of 12, 18, and 42?
   A 2
   B 3
   C 6
   D 21

4. The CN Tower in Toronto, Canada, is 1,815 feet tall. What is the prime factorization of that number?
   F $3 \times 5 \times 11^2$
   G $3 \times 7 \times 11^2$
   H $3 \times 7^2 \times 11$
   I $5 \times 7^2 \times 11$

5. The table shows the number of students in the school chorus.

School Chorus	
Students	Number
Girls	64
Boys	48

   The director plans to arrange the students in equal rows. Only girls or boys will be in each row. What is the greatest number of students that could be in each row?
   A 4
   B 8
   C 12
   D 16

6. Which list contains the common factors of 14 and 35?
   F 1, 2, 5, 7
   G 1, 2, 5, 7, 14, 35
   H 1, 7
   I 7

Holt Mathematics Grade 6

NAEP Objective Test 8.1.5.b continued

7. Which is a common multiple of 17 and 30?
   A 47
   B 51
   C 255
   D 510

8. The students at an art camp were divided into same-size groups shown in the table. All the students were able to be part of each kind of group. Which could NOT be the total number of students at the camp?

Art Camp Groups	
Activity	Number in Each Group
Oil Painting	2
Pottery	6
Weaving	4

   F 60
   G 72
   H 90
   I 108

9. Which is the greatest common factor of 20, 40, and 60?
   A 2
   B 5
   C 10
   D 20

10. Which is NOT a factor of 56?
    F 1
    G 6
    H 7
    I 8

11. Which list contains the common factors of 12 and 30?
    A 1, 2, 3, 6
    B 1, 2, 3, 4, 5, 6, 10, 12, 15, 30
    C 2, 3, 4, 6
    D 2, 3, 6

12. Eric is making fruit punch for a party. If he buys any one of these packages of cups, he could serve all the punch by filling the cups.

    Which could NOT be the amount of punch Eric is making?
    F 60 ounces
    G 120 ounces
    H 160 ounces
    I 240 ounces

NAEP Objective Test 8.1.5.d

Select the best answer for questions 1–12. Fill in the correct bubble on your answer sheet.

1. A taxi can hold 4 passengers and the driver. After one plane landed at the airport, 81 passengers wanted a taxi into the city. How many taxis are needed to take all the passengers into the city?

    A 20
    B 20.25
    C 21
    D 21.5

2. A section of the theater has been reserved for 378 people. If each row in the theater seats 18 people, about how many rows have been reserved?

    F 15 rows
    G 18 rows
    H 20 rows
    I 25 rows

3. Which of these numbers is NOT divisible by 6?

    A 296
    B 312
    C 504
    D 438

4. Karen needs to purchase paper cups for the Student Council breakfast. There are 12 cups in a packge for $1.65. If there will be 115 students and parents attending the breakfast, how many packages does Karen need to purchase?

    F 7
    G 8
    H 9
    I 10

5. How many 8-inch long ribbons can Mona cut from this spool of ribbon?

    A 2
    B 6
    C 7
    D 8

6. Gina is making bouquets at the Flower Shoppe. She has 197 roses. If she puts 8 roses in each bouquet, how many bouquets can she make?

    F 24
    G $24\frac{1}{3}$
    H $24\frac{5}{8}$
    I 25

NAEP Objective Test 8.1.5.d continued

7. A factory has made 456 small flashlights. The flashlights are packaged 18 per box. About how many boxes can be filled with the flashlights?
   - A 20
   - B 22
   - C 23
   - D 30

8. If a number is divisible by 6, it is divisible by which other number or numbers?
   - F 2
   - G 2, 3
   - H 2, 3, 4
   - I 2, 4

9. The school auditorium is 72 feet wide. Each chair is 20 inches wide. How many chairs can be put in a row across the auditorium?
   - A 43
   - B 44
   - C 45
   - D 46

10. Which of these numbers is NOT divisible by 9?
    - F 234
    - G 315
    - H 612
    - I 703

11. Michele has $15\frac{7}{8}$ cups of flour. Each cookie recipe calls for $2\frac{1}{8}$ cups of flour. About how many recipes of cookies can she make?
    - A 6
    - B 7
    - C 8
    - D 9

12. Which is the least number that is NOT a factor of 5,040?

    - F 9
    - G 10
    - H 11
    - I 13

NAEP Objective Test 8.1.5.e

Select the best answer for questions 1–12. Fill in the correct bubble on your answer sheet.

1. Which operation should be performed first to simplify this expression?

    $15 - 27 \div 9 + 8 \times 2$

    A addition
    B subtraction
    C multiplication
    D division

2. Ben simplified this expression.

    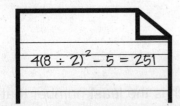

    $4(8 \div 2)^2 - 5 = 251$

    What did Ben do incorrectly?

    F He multiplied $4(8 \div 2)$ before applying the exponent.
    G He divided $(8 \div 2)$ before applying the exponent.
    H He subtracted before he divided.
    I He applied the exponent to the 2 instead of the $(8 \div 2)$.

3. Use the order of operations to simplify the expression:

    $20 + (5 - 2)^2 - 12$

    A 17
    B 23
    C 517
    D 529

4. Which operation should be performed first to simplify this expression?

    $4 + 3 - 7 - 6$?

    F multiplication
    G division
    H addition
    I subtraction

5. Use the order of operations to simplify the expression:

    $6^2 - 6 \div 6$

    A 0
    B 5
    C 6
    D 35

6. Which placement of parentheses makes the statement true?

    F $(3 + 18) \div 6 - 4 = 2$
    G $3 + (18 \div 6) - 4 = 2$
    H $(3 + 18) \div 6 - 4 = 2$
    I $3 + 18 \div (6 - 4) = 2$

NAEP Objective Test 8.1.5.e continued

7. Which placement of parentheses makes the statement true?
   A $75 + (7 \times 2^2) - 3 = 82$
   B $(75 + 7) \times 2^2 - 3 = 82$
   C $75 + (7 \times 2^2 - 3) = 82$
   D $75 + 7 \times (2^2 - 3) = 82$

8. Use the order of operations to simplify the expression:
   $19 + (4^2 + 3) - 30$
   F 1
   G 8
   H 31
   I 75

9. Use the order of operations to simplify the expression.
   $7 \times (9 - 4) + (16 \div 2)$
   A 16
   B 43
   C 58
   D 67

10. Which placement of parentheses makes the statement true?
    F $25 + 3^2 - (2^2 \times 7) = 6$
    G $(25 + 3^2 - 2^2) \times 7 = 6$
    H $25 + (3^2 - 2^2) \times 7 = 6$
    I $25 + (3^2 - 2^2 \times 7) = 6$

11. Allison simplified this expression.

    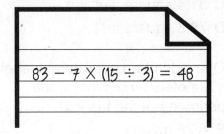
    $83 - 7 \times (15 \div 3) = 48$

    Did Allison correctly simplify the expression?
    A Yes.
    B No, the value should be $7\frac{1}{3}$.
    C No, the value should be -22.
    D No, the value should be 380.

12. Use the order of operations to simplify the expression:
    $(4^2 - 8) \times 5 + 9^2$
    F 57
    G 121
    H 244
    I 672

NAEP Objective Test 8.2.1.h

Select the best answer for questions 1–12. Fill in the correct bubble on your answer sheet.

1. A round area rug has a radius of 5 feet. How much area does the rug cover? Use 3.14 for π.
   A 15.7 square feet
   B 19.625 square feet
   C 31.4 square feet
   D 78.5 square feet

2. One side of an equilateral triangle is 4.5 meters long. What is the perimeter of the triangle?
   F 6.75 meters
   G 10.125 meters
   H 13.5 meters
   I 18 meters

3. The floor of Pablo's bedroom is a rectangle. He used 120 1-foot square tiles to completely cover it. Which of these could be the dimensions of Pablo's bedroom?
   A 8 feet × 15 feet
   B 9 feet × 12 feet
   C 10 feet × 20 feet
   D 30 feet × 30 feet

4. What is the area of a triangle with a base of 7 inches and a height of 10 inches?
   F $17\frac{1}{2}$ square inches
   G 27 square inches
   H 35 square inches
   I 70 square inches

5. A circle with a radius of 3 inches is inscribed in a square. What is the area between the circle and the square? Use 3.14 for π.

   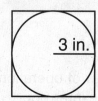

   A 6 in²
   B 7.74 in²
   C 19.26 in²
   D 36 in²

6. Marla used 64 centimeters of wood to build a square picture frame. What is the length of each side of the frame?
   F 8 cm
   G 16 cm
   H 32 cm
   I 256 cm

NAEP Objective Test 8.2.1.h continued

7. Joseph wants to make a poster shaped like a house. He has a piece of cardboard shaped like a rectangle and another piece shaped like a triangle. What will the perimeter of the poster be?

   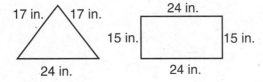

   A 71 inches
   B 88 inches
   C 112 inches
   D 136 inches.

8. Ana is sewing fringe around the edge of a circular tablecloth that has a diameter of 30 inches. How much fringe will Ana use? Use 3.14 for π.

   F 94.2 inches
   G 188.4 inches
   H 706.5 inches
   I 2,826 inches.

9. A triangle has a 24 centimeter base and a height of 15 centimeters. If the height is reduced by 5 centimeters, what will be the new area of the triangle?

   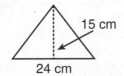

   A 120 square centimeters
   B 180 square centimeters
   C 240 square centimeters
   D 360 square centimeters

10. What is the perimeter of a 4.2 centimeter × 5.6 centimeter rectangle?

    F 9.8 centimeters
    G 19.6 centimeters
    H 23.52 centimeters
    I 44 centimeters

11. A circle has a diameter of 12 centimeters. If the diameter is reduced by 7 centimeters, what will be the area of the new circle? Use 3.14 for π.

    A 38.465 square centimeters
    B 19.625 square centimeters
    C 78.5 square centimeters
    D 153.86 square centimeters

12. Ahmad is planning a garden that is 20 feet square. In the center of the garden will be a circular pond with a diameter of 6 feet. What is the area of the garden that will be left for planting? Use 3.14 for π.

    F 286.96 square feet
    G 371.74 square feet
    H 400 square feet
    I 428.26 square feet

Name _____ Date _____ Class _____

NAEP Objective Test 8.2.1.j

Select the best answer for questions 1–12. Fill in the correct bubble on your answer sheet.

1. Each cylindrical column surrounding a monument has a radius of 3 feet and a height of 54 feet. Find the total volume of each column. Use 3.14 for π.
   A 486 cubic feet
   B 1017.36 cubic feet
   C 1526.04 cubic feet
   D 6104.16 cubic feet

2. What is the volume of a 5 foot long, 3 foot wide, and 2 foot deep rectangular toy chest?
   F 10 cubic feet   H 30 cubic feet
   G 15 cubic feet   I   60 cubic feet

3. What is the surface area of the cube?

   7 cm

   A 84 square centimeters
   B 147 square centimeters
   C 294 square centimeters
   D 343 square centimeters

4. A can has a diameter of 5 inches and a height of 7 inches. What is the area of a label that goes all around the can? Use 3.14 for π.
   F 109.9 square inches
   G 137.375 square inches
   H 149.15 square inches
   I   219.8 square inches

5. Which of the following cylinders has the greatest volume?

   A 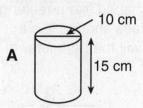 10 cm / 15 cm

   B 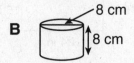 8 cm / 8 cm

   C 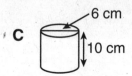 6 cm / 10 cm

   D 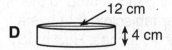 12 cm / 4 cm

6. Kara has 1,000 square inches of wrapping paper. How many of the following gift boxes can she wrap?

   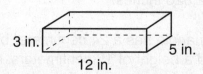
   3 in.   5 in.
   12 in.

   F 2 boxes
   G 3 boxes
   H 4 boxes
   I   5 boxes

NAEP Objective Test 8.2.1.j continued

7. The volume of a cube is 27 cubic feet. What is the length of one edge of the cube?

   A 3 feet
   B 4.5 feet
   C 9 feet
   D 19,683 feet

8. Julius wants to cover the surface of the pyramid shown with brown paper for a social studies project. How many square centimeters of paper will he need?

   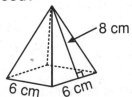

   F 96 square centimeters
   G 132 square centimeters
   H 144 square centimeters
   I 228 square centimeters

9. Maria is wrapping a present. The shipping box measures 10 inches by 5 inches by 5 inches. A small gift box inside the shipping box measures 2 inches by 2 inches by 2 inches. How much space is left inside the shipping box for packaging material?

   A 8 cubic inches
   B 242 cubic inches
   C 250 cubic inches
   D 258 cubic inches

10. What is the surface area of this figure made by gluing three cubes together?

    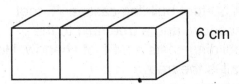

    F 216 square centimeters
    G 252 square centimeters
    H 504 square centimeters
    I 648 square centimeters

11. What is the volume of this cabinet?

    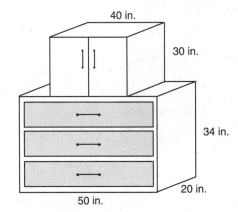

    A 58,000 cubic inches
    B 64,000 cubic inches
    C 70,000 cubic inches
    D 115,200 cubic inches

12. The surface area of a cube is 24 square centimeters. What is the length of one edge of the cube?

    F 2 centimeters
    G 4 centimeters
    H 8 centimeters
    I 2688 centimeters

NAEP Objective Test 8.2.1.k

Select the best answer for questions 1–12. Fill in the correct bubble on your answer sheet.

1. A 30-foot building casts a 45-foot shadow, and a tree next to the building casts a 24-foot shadow. How tall is the tree?
   A 16 feet
   B 39 feet
   C 36 feet
   D 51 feet

2. A 4-foot sign casts a shadow 7 feet long. How long a shadow does the building behind it cast?

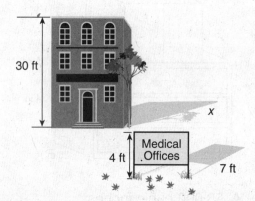

   F 17.1 feet
   G 33 feet
   H 52.5 feet
   I 210 feet

3. A 10-foot lamppost stands next to a 24-foot pine tree. When the pine tree casts a shadow 30 feet long, how long a shadow does the lamppost cast?
   A 8 feet   C 12.5 feet
   B 10.6 feet   D 16 feet

4. Andrew planted a young maple tree 8 feet tall that casts a shadow 12 feet long. At the same time, an elm tree in Andrew's yard casts a shadow 42 feet long. What is the height of the elm tree?
   F 2.3 feet   H 38 feet
   G 28 feet   I 63 feet

5. Terrence is standing on a sidewalk next to a lamppost. The lamppost is 9 feet tall and casts a 3-foot shadow. Terrence is 6 feet tall. How long is his shadow?
   A 2 feet
   B 4.5 feet
   C 12 feet
   D 18 feet

6. Jen casts a shadow 2 meters long at the same time that the tree she is standing by casts a shadow 15 meters long. If the tree is 9 meters tall, how tall is Jen?
   F 1.2 meters
   G 3.3 meters
   H 8 meters
   I 67.5 meters

NAEP Objective Test 8.2.1.k continued

7. A sign casts a shadow that is 30 feet long. A person standing nearby casts a shadow 11 feet long. What is the height of the sign?

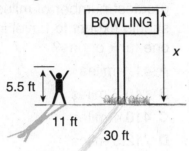

   A 2 feet
   B 82.5 feet
   C 15 feet
   D 60 feet

8. The oldest rose bush in Julia's garden is 35 inches tall and casts a shadow 56 inches long. The newest rose bush in the garden is only 10 inches tall. How long a shadow does it cast?

   F 6.25 inches
   G 16 inches
   H 31 inches
   I 196 inches

9. The flagpole at Jaime's school casts a shadow 40 meters long. If Jaime is 1.25 meters tall and casts a shadow of 8 meters at the same time of day, what is the height of the flagpole?

   A 5.5 meters
   B 6.25 meters
   C 10 meters
   D 32 meters

10. Ryan is 1.6 meters tall and casts a shadow 3 meters long. He is visiting the Space Needle in Seattle, WA, which is 184 meters tall. How long is the Space Needle's shadow?

    F 61.3 meters
    G 98.1 meters
    H 185.4 meters
    I 345 meters

11. Scott is 6 feet tall and casts a shadow 9 feet long. He is standing beside the tent he just pitched which casts a shadow 7.5 feet long. What is the height of the tent?

    A 4.5 feet
    B 5 feet
    C 7.2 feet
    D 11.25 feet

12. Tamara is 4.5 feet tall and casts a shadow 6 feet long. She is standing beside her house that is 20 feet tall. How long a shadow does her house cast?

    F 1.29 feet
    G 15.75 feet
    H 21.5 feet
    I 26.7 feet

NAEP Objective Test 8.2.1.I

Select the best answer for questions 1–12. Fill in the correct bubble on your answer sheet.

1. A grocery store has a sale on red grapes of 5 pounds for $6. What is the unit price of the grapes?
   A $0.38 per pound
   B $0.83 per pound
   C $1.11 per pound
   D $1.20 per pound

2. Mr. Ortez is driving 250 miles from Dallas to Houston. The fuel efficiency of his car is 32 miles per gallon. To the nearest tenth, how many gallons of gas will Mr. Ortez use to make this trip?
   F 7.5 gallons
   G 7.8 gallons
   H 7.9 gallons
   I 8.0 gallons

3. A store sold 2,250 video games in June. Which rate describes the store's video game sales?
   A 75 games per day
   B 450 games per week
   C 9,000 games per week
   D 67,500 games per day

4. There are 9 calories per gram of fat in food. How many calories come from fat in a serving of lasagna that contains 25 grams of fat?
   F 2.77 calories
   G 34 calories
   H 225 calories
   I 277 calories

5. The gas tank in Ms. Miller's car holds 17.1 gallons of gas and has a fuel efficiency of 24 miles per gallon when driving in the city. What is the greatest number of miles Ms. Miller should expect to travel in the city on one tank of gas?
   A 41.1 miles
   B 140.3 miles
   C 410.4 miles
   D 712.5 miles

6. The table shows the population and land area of some states in 2000. Which of these states has the greatest population density (people per square mile)?

State	Population (thousands)	Area (thousands of square miles)
California	33,872	156
New Jersey	8414	7
Rhode Island	1048	1
Texas	20,852	262

   F California
   G New Jersey
   H Rhode Island
   I Texas

NAEP Objective Test 8.2.1.I continued

7. If the unit price of a gallon of milk is $0.0257 per ounce, what is the cost of a gallon of milk? (Hint: 1 gallon = 128 fluid ounces)
   A $1.31
   B $2.07
   C $2.57
   D $3.29

8. A biscuit recipe calls for $\frac{3}{4}$ cup of milk per 2 cups of flour. How much milk is needed if 6 cups of flour are used?
   F $2\frac{1}{4}$ cups
   G 3 cups
   H 4 cups
   I $4\frac{1}{2}$ cups

9. The table shows the cost of different size brands of cereal at one grocery store.

Brand	Size (ounces)	Price
A	14	$2.89
B	15	$3.19
C	21	$2.79
D	22	$3.99

   Which brand of cereal is the best buy?
   A Brand A
   B Brand B
   C Brand C
   D Brand D

10. A cheetah's maximum speed over a 0.25 mile distance was found to be 70 miles per hour. To the nearest tenth, how long did it take the cheetah to run 0.25 mile?
    F 0.2 minute
    G 0.3 minute
    H 2.8 minutes
    I 280.0 minutes

11. Which of the foods listed in the table has the highest amount of protein per ounce?

Food	Quantity (ounces)	Protein (grams)
Cooked rice	5	2.7
Kidney beans	4	7.6
Low fat yogurt	6	8.9
Whole milk	8	8

    A cooked rice
    B kidney beans
    C low fat yogurt
    D whole milk

12. Sherry swims 2.5 miles every day in 35 minutes. If she swims at the same rate, how long will it take her to swim 15 miles?
    F 2.25 hours
    G 180 minutes
    H 3.5 hours
    I 210 hours

NAEP Objective Test 8.2.2.b

Select the best answer for questions 1–12. Fill in the correct bubble on your answer sheet.

1. How many 8 inch long ribbons can Mona make from a 4 foot piece of material?
   A 6 ribbons
   B 8 ribbons
   C 10 ribbons
   D 12 ribbons

2. Which measurement is equivalent to 2.8 liters?
   F 0.28 milliliter
   G 28 milliliters
   H 280 milliliters
   I 2,800 milliliters

3. What is the weight in pounds of a $1\frac{1}{2}$ ton pickup truck?
   A 1,500 pounds
   B 2,400 pounds
   C 2,500 pounds
   D 3,000 pounds

4. Vanessa is cutting 200 centimeter long pieces of string to attach to balloons. How many pieces can she cut from 40 meters of string?
   F 20 pieces
   G 50 pieces
   H 200 pieces
   I 500 pieces

5. Which of the following is NOT equivalent to $4\frac{1}{2}$ yards?
   A 12 feet 18 inches
   B $13\frac{1}{2}$ feet
   C 150 inches
   D 162 inches

6. Which of these trails at one state park is the longest?

Trail	Length
Deer Meadow	4.07 kilometers
Lakeside	4.5 kilometers
Overlook	4,008 meters
Wildflower	4 kilometers 600 meters

   F Deer Meadow
   G Lakeside
   H Overlook
   I Wildflower

NAEP Objective Test 8.2.2.b continued

7. Which measurement is equivalent to 55 milligrams?

   A 0.055 gram
   B 0.55 gram
   C 5,500 grams
   D 55,000 grams

8. Which bottle will hold the greatest amount?

   F $1\frac{1}{2}$ gallon bottle
   G 7 quart bottle
   H 10 pint bottle
   I 144 fluid ounce bottle

9. Enoch's thermos can hold 12 fluid ounces of liquid. How many times can he fill his thermos from a half gallon bottle of juice?

   A 2 times
   B 4 times
   C 5 times
   D 6 times

10. Roberto's math book is 3.8 centimeters thick. Which is another way to describe the thickness of Roberto's book?

    F 0.38 millimeter
    G 38 millimeters
    H 308 millimeters
    I 3,800 millimeters

11. What is the cost for Charlotte to ship a package that weighs 50 ounces?

Wrap and Ship Service	
Up to 1 pound	$7.25
Each additional pound or fraction of a pound	$0.50

    A $1.75
    B $7.75
    C $8.75
    D $9.25

12. Each of Dana's footsteps is $1\frac{1}{2}$ feet long. How many steps would Dana have to take to walk 1 mile?

    F 667 steps
    G 1,500 steps
    H 3,520 steps
    I 7,920 steps

NAEP Objective Test 8.2.2.d

Select the best answer for questions 1–12. Fill in the correct bubble on your answer sheet.

1. Which unit of measure are the eighth graders' heights given in?

Eighth Grader	Height
Ted	165
Suzanna	158
Franklin	164
Mia	160

   A centimeters
   B feet
   C inches
   D millimeters

2. Which is most likely the area covered by a large farm?

   F 6 square centimeters
   G 6 square kilometers
   H 6 square meters
   I 6 square millimeters

3. Which unit of measure would be best to find the room inside the trailer of a semi truck?

   A cubic centimeters
   B cubic inches
   C cubic miles
   D cubic yards

4. Jane made this scale drawing of a bug. She labeled its actual length. Which unit of measure is missing from the label?

   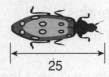

   F centimeters
   G inches
   H millimeters
   I yards

5. Which unit of measure would be best to use to measure the base of a flower pot?

   A square centimeters
   B square kilometers
   C square meters
   D square millimeters

6. Patrick said that the number of units from his house to school is 10. Which of these units did Patrick most likely use?

   F centimeters
   G kilometers
   H meters
   I millimeters

NAEP Objective Test 8.2.2.d continued

7. Ilia is using this box to send a pottery vase to her aunt. The numbers on the box indicate the dimensions of the box. What is the width of this box?

   A 12 feet
   B 12 inches
   C 12 miles
   D 12 yards

   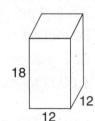

8. Matt measured several features of his bedroom. However, when he recorded his findings, he forgot to record the units of measure he used. One measurement he recorded is 34. What feature's measurement would this most likely be?

   F the area of the window in square feet
   G the height of the room in feet
   H the length of the room in yards
   I the width of the door in inches

9. A can is 1 foot tall. Its diameter is a little shorter than its height. A label goes completely around it. Which is the best estimate for the area of the label?

   A 3 square feet
   B 30 square feet
   C 30 square inches
   D 300 square inches

10. What unit is missing from the sign on the door of this refrigerator that is for sale?

    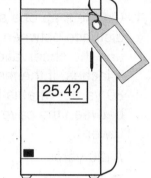

    F cubic feet
    G cubic inches
    H cubic miles
    I cubic yards

11. Which of the following is NOT likely a dimension of a cereal box?

    A 15 millimeters
    B 19 centimeters
    C 23 centimeters
    D 57 millimeters

12. This gift box can be used to hold a ring. Which is most likely the missing number in the label?

    F 0.16
    G 16
    H 160
    I 1,600

NAEP Objective Test 8.2.2.f

Select the best answer for questions 1–12. Fill in the correct bubble on your answer sheet.

1. On the map of a state park, the distance between the cave entrance and the observation tower is 4 inches. If the map scale is 1 inch = 80 yards, what is the actual distance between the cave entrance and the tower?
   A 20 yards
   B 40 yards
   C 84 yards
   D 320 yards

2. The scale on a map is 2 inches = 15 miles. If two cities are 75 miles apart, how far apart will they appear on the map?
   F 2.5 inches
   G 5 inches
   H 10 inches
   I 37.5 inches

3. How far will Alex travel from home to school if he stops at the library on his way?

   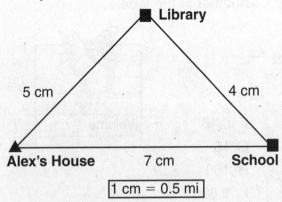

   A 1.8 miles
   B 3.5 miles
   C 4.5 miles
   D 9 miles

4. The scale factor for a model ship is 3 inches = 20 feet. If the length of the ship is 150 feet, what is the length of the model?
   F 7.5 inches
   G 50 inches
   H 22.5 inches
   I 1,000 inches

5. The scale factor for a model plane is 2 inches = 25 feet. If the length of the model is 8 inches, how long is the plane?
   A 100 inches
   B 8 feet
   C 80 feet
   D 100 feet

6. Jorge caught a fish 9 inches long and made this scale drawing of it. What scale did he use?

   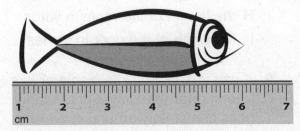

   F 1 centimeter = 0.67 inch
   G 1 centimeter = 1.5 inches
   H 1 inch = 1.5 centimeters
   I 1 inch = 3 centimeters

NAEP Objective Test 8.2.2.f continued

7. Alicia used the scale 1 inch = 1 foot to make a scale drawing of her bedroom. She drew her dresser inches $1\frac{3}{4} \times 3\frac{1}{2}$ inches. What are the actual dimensions of her dresser?

   A $12\frac{3}{4}$ inches × $36\frac{1}{2}$ inches
   B $13\frac{3}{4}$ inches × $15\frac{1}{2}$ inches
   C 21 inches × 42 inches
   D 30 inches × 56 inches

8. Kari wants to make a scale drawing of her family on a sheet of paper 11 inches long. Kari's father is 6.25 feet tall and is the tallest family member. Which scale could Kari use for her drawing?

   F 1 inch = 0.5 foot
   G 2 inches = 1 foot
   H 3 inches = 2 feet
   I 4 inches = 1.5 feet

9. The scale 1 inch = 6 feet was used to draw the plans for a house. If the drawing is 10.5 inches wide, how wide is the house?

   A 10.5 feet
   B 17.5 feet
   C 60.5 feet
   D 63 feet

10. The scale on a map of California is 3 inches = 40 miles. Which of the cities listed in the table is about 9 inches from Los Angeles on the map?

City	Distance from Los Angeles (in miles)
Monterey	327
San Diego	127
San Francisco	387
Santa Barbara	91

    F Monterey
    G San Diego
    H San Francisco
    I Santa Barbara

11. Jermaine found that a Tyrannosaurus Rex could be 20 feet tall and 49 feet long. If he wants to make a scale drawing of this dinosaur using the scale 2 inches = 5 feet, which of these size sheets of paper could he use for his drawing?

    A 3 inches × 11 inches
    B 5 inches × 8 inches
    C 6 inches × 9 inches
    D 8 inches × 30 inches

12. Mr. Quick made a scale drawing of a rectangular dog pen 3 centimeters by 4 centimeters. If he used the scale 1 centimeter = 6 feet, what is the actual perimeter of the pen?

    F 14 centimeters
    G 42 feet
    H 84 feet
    I 504 feet

NAEP Objective Test 8.3.1.b

Select the best answer for questions 1–12. Fill in the correct bubble on your answer sheet.

1. Which solid has exactly two parallel faces?
   A cube
   B cone
   C sphere
   D cylinder

2. Which figure always has four congruent sides?
   F rectangle
   G parallelogram
   H rhombus
   I trapezoid

3. Which figure has two endpoints?

   A

   B

   C

   D

4. Which figure has exactly eight vertices?
   F pentagon
   G hexagon
   H octagon
   I polygon

5. Which type of triangle has two 70° angles?
   A equilateral
   B obtuse
   C isosceles
   D scalene

6. Which figure CANNOT have opposite sides parallel?
   F triangle
   G quadrilateral
   H pentagon
   I hexagon

NAEP Objective Test 8.3.1.b continued

7. Which solid has 6 edges?

A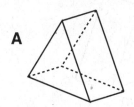

B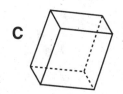

C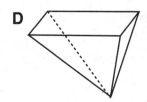

D

8. Which figure is NOT a quadrilateral?
   F parallelogram
   G pentagon
   H rhombus
   I trapezoid

9. Which type of triangle CANNOT have two 4-centimeter sides?
   A right
   B isosceles
   C scalene
   D equilateral

10. Which angle measures less than 90°?
    F obtuse
    G right
    H acute
    I scalene

11. Which types of lines intersect in a right angle?
    A parallel
    B perpendicular
    C intersecting, but not congruent
    D intersecting, but not perpendicular

12. Which figure has five triangular faces?
    F triangular prism
    G pentagonal pyramid
    H triangular pyramid
    I pentagonal prism

Name _____ Date _____ Class _____

NAEP Objective Test 8.3.1.c

Select the best answer for questions 1–12. Fill in the correct bubble on your answer sheet.

1. Identify this shape.

   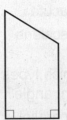

   A pentagon
   B parallelogram
   C rhombus
   D trapezoid

2. Quadrilateral WXYZ is a rectangle. Which term describes ∠WYX?

   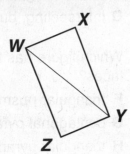

   F acute
   G obtuse
   H diagonal
   I right

3. Which term describes sides EA and AB?

   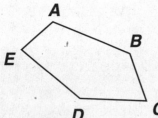

   A obtuse
   B adjacent
   C congruent
   D perpendicular

4. Identify this shape.

   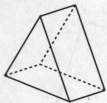

   F triangular prism
   G congruent triangles
   H triangular pyramid
   I rectangular pyramid

5. Which describes lines PQ and XY?

   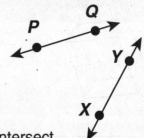

   A never intersect
   B perpendicular
   C intersect in one point
   D intersect in two points

6. Which term best describes this shape?

   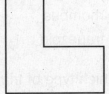

   F regular
   G parallel
   H hexagon
   I rectangle

Copyright © by Holt, Rinehart and Winston.
All rights reserved.

Holt Mathematics Grade 6

NAEP Objective Test 8.3.1.c continued

7. Identify this shape.

   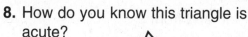

   A cube
   B parallelogram
   C prism
   D pyramid

8. How do you know this triangle is acute?

   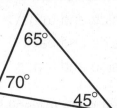

   F It has one acute angle.
   G It has three acute angles.
   H All the angles are different.
   I None of the angles equals 90°.

9. Which line segment is a height for this triangle?

   A $\overline{SO}$
   B $\overline{SQ}$
   C $\overline{TR}$
   D $\overline{RO}$

   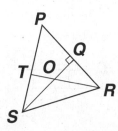

10. The acute angles of this rhombus measure 60°. Which term describes △HJL?

    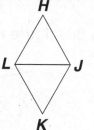

    F isosceles
    G equilateral
    H quadrilateral
    I scalene

11. How many edges does this solid have?

    A 5
    B 6
    C 10
    D 15

    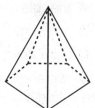

12. What does this figure show?

    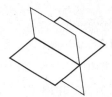

    F two parallel planes
    G two perpendicular planes
    H four intersecting planes
    I four perpendicular planes

Name _____ Date _____ Class _____

NAEP Objective Test 8.3.1.e

Select the best answer for questions 1–12. Fill in the correct bubble on your answer sheet.

Use this figure for questions 1–3.

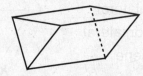

1. What does the solid look like from the top?
   A triangle
   B rectangle
   C trapezoid
   D rectangle with line down the middle

2. Which two views of the solid are the same?
   F front and back
   G top and right side
   H top and bottom
   I bottom and right side

3. Which term best describes the front view?
   A isosceles
   B equilateral
   C congruent
   D parallel

Use this figure for questions 4–6.

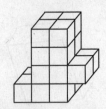

4. Which two views of the solid are the same?
   F top and front
   G side and front
   H side and back
   I side and top

5. What does the solid look like from the top?

   A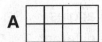

   B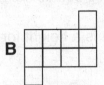

   C

   D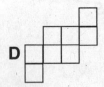

6. How many cubes are used to build the solid?
   F 10
   G 12
   H 20
   I 24

Name _____ Date _____ Class _____

NAEP Objective Test 8.3.1.e continued

7. What is the top view of this solid?

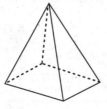

   A square
   B rectangle
   C triangle
   D four triangles

8. Which views of this cylinder are NOT the same?

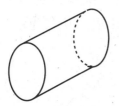

   F front and back
   G top and right side
   H front and right side
   I right and left sides

9. Which solid CANNOT have a rectangle as one of its views?

   A prism
   B sphere
   C pyramid
   D cylinder

Use this figure for questions 10–12.

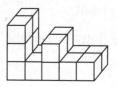

10. Which shows the right side view?

   F

   G

   H

   I

11. Which describes the top view?

   A a 2-by-3 rectangle
   B a 2-by-5 rectangle
   C a 2-by-6 rectangle
   D a non-rectangular polygon

12. If you remove the 2 top cubes, which views do NOT change?

   F top only
   G right side only
   H top and right side
   I None of the views change.

NAEP Objective Test 8.3.2.a

Select the best answer for questions 1–12. Fill in the correct bubble on your answer sheet.

1. How many lines of symmetry does this figure have?

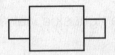

   A 0
   B 1
   C 2
   D 4

2. Which figure will look the same after a 90° turn clockwise?

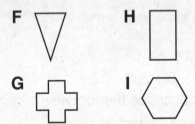

3. How many lines of symmetry does a square have?
   A 1
   B 2
   C 4
   D 8

4. Which line is a line of symmetry for this trapezoid?

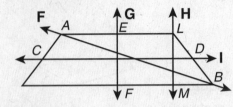

   F AB
   G CD
   H EF
   I LM

5. Which type of triangle always has exactly one line of symmetry?
   A right
   B isosceles
   C equilateral
   D scalene

6. In the figure below, why isn't the line a line of symmetry?

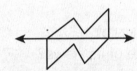

   F The two parts of the figure are not congruent.
   G The two parts of the figure are not regular.
   H The line is not a diagonal of the figure.
   I The bottom part of the figure is not a reflection of the top part.

NAEP Objective Test 8.3.2.a continued

7. Which quadrilateral could have 0 lines of symmetry?
   A square
   B rectangle
   C parallelogram
   D rhombus

8. Which figure does NOT have a line of symmetry?

   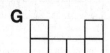

9. Which design has BOTH rotational symmetry and line symmetry?

10. How many lines of symmetry does this design have?

    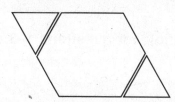

    F 0
    G 2
    H 3
    I 6

11. Which figure does NOT look the same after a $\frac{1}{3}$ turn clockwise?

    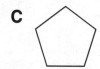

12. Which shape has an infinite number of lines of symmetry?
    F circle
    G square
    H equilateral triangle
    I any regular polygon

Name _____ Date _____ Class _____

NAEP Objective Test 8.3.2.c

Select the best answer for questions 1–12. Fill in the correct bubble on your answer sheet.

Use this triangle for questions 1–3.

1. What does the triangle look like after a reflection across a horizontal line?

   A   C

   B   D

2. What does the triangle look like after a $\frac{1}{4}$ turn rotation counterclockwise?

   F   H

   G   I

3. Under which transformation will the triangle have the same shape?

   A any of these transformations below
   B 180° rotation
   C 90° rotation clockwise
   D reflection across a vertical line

Use this figure for questions 4–6.

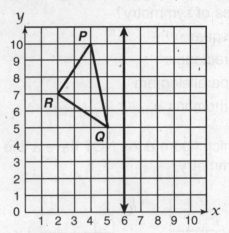

4. What are the coordinates of point R after △PQR is translated 2 units down and 5 units to the right?

   F (2, 4)
   G (6, 5)
   H (6, 7)
   I (7, 5)

5. Where will point P be after the triangle is reflected across the vertical line?

   A (7, 10)
   B (8, 10)
   C (10, 7)
   D (10, 8)

6. Which translation will move point Q to the location (3, 1)?

   F 2 left, 3 down
   G 2 left, 4 down
   H 3 right, 1 down
   I 4 left, 2 down

NAEP Objective Test 8.3.2.c continued

Use this triangle for questions 7–8.

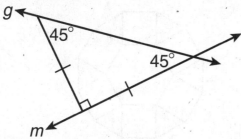

7. What figure do you get if you flip the right triangle across line *g*?

   A a square
   B a rectangle
   C a parallelogram
   D a larger right triangle

8. What figure do you get if you flip the right triangle across line *m*?

   F a square
   G a rectangle
   H a parallelogram
   I a larger right triangle

9. Which figure looks the same after it is rotated 60° clockwise?

   A   C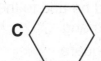

   B   D

Use this figure for questions 10–12.

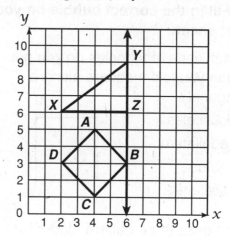

10. Triangle *XYZ* is rotated $\frac{1}{4}$ turn clockwise about point *Z*. What is the new location of vertex *X*?

    F (2, 6)   H (6, 9)
    G (6, 6)   I (6, 10)

11. Which translation will make vertex *X* coincide with vertex *B*?

    A 3 right, 4 down
    B 4 right, 3 down
    C 5 right, 3 down
    D 5 right, 4 down

12. Square *ABCD* is flipped across the line of reflection. Which one of these changes?

    F the area
    G the angle measures
    H the coordinates of vertex *B*
    I the coordinates of vertex *D*

NAEP Objective Test 8.3.2.d

Select the best answer for questions 1–12. Fill in the correct bubble on your answer sheet.

1. What kinds of triangles do you get when you fold a square along a diagonal?

   A isosceles
   B equilateral
   C obtuse
   D scalene

2. Tom used three pattern blocks to make this polygon. Which term describes the figure Tom made?

   F hexagon
   G pentagon
   H octagon
   I quadrilateral

3. Which shape CANNOT be made by joining two isosceles right triangles?

   A   C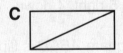

   B    D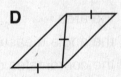

Use this design for questions 4–6.

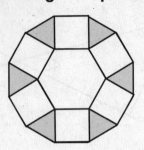

4. What shapes are used to make the design?

   F square, triangle
   G square, hexagon
   H square, triangle, octagon
   I square, triangle, hexagon

5. Mandy made the design and then removed the triangles. How many sides does her new design have?

   A 6   C 18
   B 12   D 24

6. Ed continued the pattern in the design by joining a hexagon to the outside edge of each square. What should he add to the outside edge of each triangle?

   F a square
   G a pentagon
   H a hexagon
   I an octagon

NAEP Objective Test 8.3.2.d continued

7. Phillip's little sister used these four blocks to build a tower. Which block has to be on top?

 A   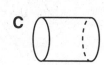 C

 B   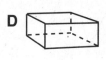 D

8. If you replace the hexagon with triangles, how many small triangles will there be in all?

   F 3
   G 4
   H 9
   I 12

   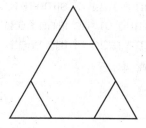

9. Which line segment will divide the shape into two parallelograms?

   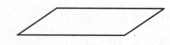

   A vertical
   B horizontal
   C diagonal starting at top left
   D diagonal starting at bottom left

10. How many small trapezoids are used to make this larger similar trapezoid?

    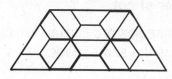

    F 9   H 12
    G 10   I 16

11. What do you get if you join one side of the square to one leg of the right triangle?

    A a large right triangle
    B a trapezoid
    C a pentagon
    D a trapezoid or a pentagon

12. A large rectangluar shipping box is completely filled with smaller boxes so that there are no holes and no extra space. Which of these CANNOT be the shape of the small boxes?

     F   H

     G   I

NAEP Objective Test 8.3.2.e

Select the best answer for questions 1–12. Fill in the correct bubble on your answer sheet.

1. These triangles are similar. Which angle has the same measure as ∠N?

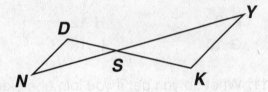

   A ∠K
   B ∠Y
   C ∠DSN
   D ∠YSK

2. In which figure can you draw a diagonal and get two congruent right triangles?

   F a rectangle
   G a right triangle
   H any parallelogram
   I a rhombus that does not have 90° angles

3. Which of the following are always similar?

   A two rectangles
   B two parallelogram
   C two right triangles
   D two squares

4. Which polygon is NOT congruent to this shape?

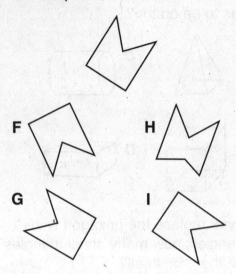

5. In a pair of similar rectangles, the ratio of the lengths is 4 : 5. What is the ratio of the widths?

   A 4 : 5
   B 5 : 4
   C 4 : 9
   D 9 : 4

6. What do two congruent angles always have?

   F the same endpoints
   G rays of equal length
   H the same measure
   I the same vertex

NAEP Objective Test 8.3.2.e continued

7. Which term describes the corresponding angles in a pair of similar figures?

A similar

B congruent

C proportional

D parallel

8. Which set of side lengths will make a triangle similar to one with sides of 6 centimeters, 6 centimeters, and 6 centimeters?

F 2 cm, 2 cm, 2 cm

G 2 cm, 3 cm, 6 cm

H 3 cm, 3 cm, 6 cm

I 3 cm, 6 cm, 6 cm

9. Which pair of figures is similar?

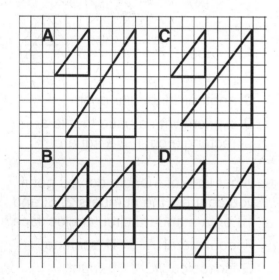

Use this pair of similar triangles for questions 10–12. The ratio of corresponding sides is 2 : 1.

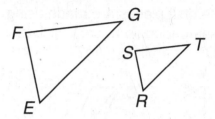

10. Which side in △RST corresponds to side EF?

F side FG

G side EG

H side ST

I side RS

11. If side RT is 5 centimeters, what is the length of side EG?

A 2.5 cm

B 5 cm

C 10 cm

D 15 cm

12. Which angle measure could you use to find the measure of ∠F?

F m∠E

G m∠G

H m∠R

I m∠S

NAEP Objective Test 8.3.3.b

Select the best answer for questions 1–12. Fill in the correct bubble on your answer sheet.

1. These quilt pieces are made using the same triangle twice. Which piece is NOT a parallelogram?

   A   C

   B   D

2. Olivia made this rug design by dividing the sides of the rectangle in fourths. Which polygon CANNOT be found in her design?

   F rhombus
   G parallelogram
   H right triangle
   I scalene triangle

3. Two lines intersect in 1 point. What is the maximum number of intersection points for 4 lines?

   A 3
   B 4
   C 6
   D 10

   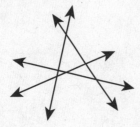

4. Phil is building a fence to close in a garden. This figure shows the fence he has built so far.

   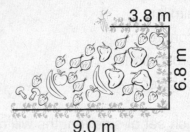

   Which of these is NOT a possible shape for Phil's finished garden?

   F rectangle   H square
   G trapezoid   I pentagon

5. What do you get if you glue the gray faces of these identical triangular prisms together?

   A a cube
   B pyramid
   C another triangular prism
   D a prism with a base that is a rectangle but not a square

6. What kind of triangle do you get if you connect the numbers 2, 6, and 10 on a standard circular clockface?

   F equilateral   H scalene
   G isosceles   I right

Holt Mathematics Grade 6

NAEP Objective Test 8.3.3.b continued

Use this figure for questions 7–8. Side *KL* is one-half as long as *KH*.

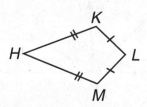

7. Which points can you connect to get two congruent triangles?
   A *K* and *M*
   B *K* and *L*
   C *L* and *M*
   D *L* and *H*

8. Which equals the perimeter?
   F 2 · *KL*
   G 3 · *KH*
   H 4 · *KH*
   I 4 · *KL*

9. Which of these angles is larger?

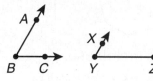

   A *ABC*
   B *XYZ*
   C They are the same size.
   D There is not enough information to answer the question.

10. A revolving restaurant makes a complete rotation in 1 hour. How many minutes does it take the restaurant to rotate 90°?
    F 15
    G 30
    H 45
    I 90

11. Greg found the center of a circular table cloth by folding it in half two different ways. Why does this strategy work?

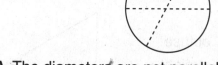

    A The diameters are not parallel.
    B The diameters are not perpendicular.
    C Diameters are lines of symmetry.
    D Any two diameters intersect in the center of the circle.

12. Carla made this cardboard pattern to build a square pyramid. What's wrong with Carla's pattern?

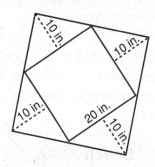

    F The four triangles aren't equilateral.
    G The base in the center should be a triangle.
    H The heights of the triangles must be greater than 10 inches.
    I The heights of the triangles must equal the perimeter of the square.

Name _____ Date _____ Class _____

NAEP Objective Test 8.3.3.f

Select the best answer for questions 1–12. Fill in the correct bubble on your answer sheet.

1. Which term describes each angle in this equilateral triangle?

   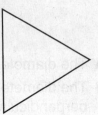

   A obtuse
   B straight
   C acute
   D right

2. Which property is NOT necessarily true of a rhombus?
   F It has four right angles.
   G It has four congruent sides.
   H Opposite sides are parallel.
   I Adjacent angles are supplementary.

3. The acute angles in this parllelogram each measure 45°. What is the measure of one of the obtuse angles?

   A 90°
   B 135°
   C 225°
   D 270°

4. Which two angles are NOT supplementary?

   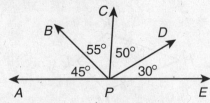

   F ∠APC and ∠CPE
   G ∠DPE and ∠DPA
   H ∠DPC and ∠CPB
   I ∠EPB and ∠BPA

5. Which property is true of all regular polygons?

   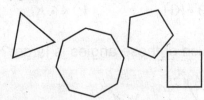

   A All angles are obtuse.
   B All angles are congruent.
   C Opposite sides are parallel.
   D Adjacent sides are perpendicular.

6. Which of these figures is NOT a parallelogram?
   F square
   G rhombus
   H rectangle
   I trapezoid

NAEP Objective Test 8.3.3.f continued

7. The corners of the square are on the circle. Which two lengths are congruent?

   A the side of the square and the diameter of the circle
   B the side of the square and the radius of the circle
   C a diagonal of the square and the diameter of the circle
   D a diagonal of the square and the radius of the circle

8. Terry drew a 15° angle and then used it as part of a right triangle. What are the measures of the other two angles in Terry's triangle?
   F 15°, 75°   H 75°, 75°
   G 15°, 90°   I 75°, 90°

9. Which term describes the adjacent angles of any parallelogram?

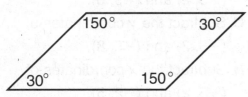

   A congruent
   B obtuse
   C complementary
   D supplementary

10. A triangle with two congruent sides cannot be
    F scalene.
    G right.
    H isosceles.
    I equilateral.

11. Alicia is making a banner in the shape of an isosceles triangle. If the smallest angle measures 20°, what is the measure of each of the other angles?

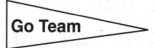

    A 20°
    B 80°
    C 120°
    D 160°

12. Two intersecting lines form four angles. Which term describes each pair of opposite angles?

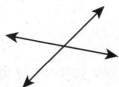

    F acute
    G congruent
    H complementary
    I supplementary

NAEP Objective Test 8.3.4.d

Select the best answer for questions 1–12. Fill in the correct bubble on your answer sheet.

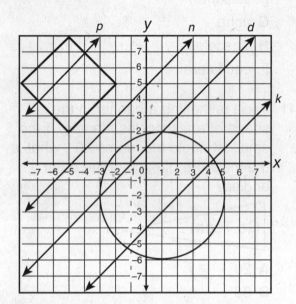

Use this figure for questions 1–6.

1. What are the coordinates of the bottom vertex of the square?
   A (2, −5)
   B (−2, 5)
   C (5, −2)
   D (−5, 2)

2. Which two vertices of the square have the same x-coordinate?
   F top and right
   G bottom and right
   H top and bottom
   I left and right

3. Which of the following locations is outside of the square?
   A (−8, 5)
   B (−6, 6)
   C (−4, 7)
   D (−4, 2)

4. Which line contains the ordered pair (−7, −6)?
   F line p
   G line n
   H line d
   I line k

5. Which of these points is the center of the circle?
   A (1, −2)
   B (2, −2)
   C (−2, −2)
   D (2, −3)

6. How can you find the length of a diagonal of the square?
   F Subtract the x-coordinates of (−5, 2) and (−5, 8).
   G Subtract the y-coordinates of (−5, 2) and (−5, 8).
   H Subtract the x-coordinates of (−5, 2) and (−2, 5).
   I Subtract the y-coordinates of (−5, 2) and (−2, 5).

NAEP Objective Test 8.3.4.d continued

Use this figure for questions 7–9.

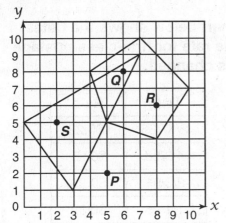

7. Which point is located at the coordinates (6, 8)?
   A Point P
   B Point Q
   C Point R
   D Point S

8. Which of these ordered pairs is NOT a vertex of the pentagon?
   F (4, 9)
   G (5, 5)
   H (7, 10)
   I (8, 4)

9. Which ordered pair represents a point located inside the pentagon and outside the triangle?
   A (2, 5)
   B (6, 8)
   C (8, 4)
   D (8, 6)

Use this figure for questions 10–12.

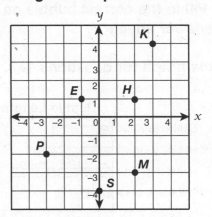

10. Where is point P?
    F (−2, -3)
    G (−3, 2)
    H (−3, −2)
    I None of these

11. Which point is at (2, −3)?
    A point P
    B point E
    C point H
    D None of these

12. Where is a point that has 0 as the x-coordinate?
    F x-axis
    G y-axis
    H quadrant III
    I quadrant IV

Name _____ Date _____ Class _____

NAEP Objective Test 8.4.1.a

Select the best answer for questions 1–12. Fill in the correct bubble on your answer sheet.

Use this graph for questions 1–3.

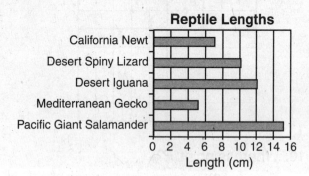

1. Which reptile is twice as long as the Mediteranean Gecko?
   A California Newt
   B Desert Spiny Lizard
   C Desert Iguana
   D Pacific Giant Salamander

2. What is the length of the middle sized reptile in this group?
   F 7 cm
   G 10 cm
   H 12 cm
   I 14 cm

3. How many of these reptiles are less than 13 centimeters long?
   A 2
   B 3
   C 4
   D 5

Use this chart for questions 4–6.

The chart below shows the dates that the rate for first-class postage stamps changed.

Date	Rate
2/17/85	$0.22
4/3/88	$0.25
2/3/91	$0.29
1/1/95	$0.32
1/10/99	$0.33
1/7/01	$0.34
6/30/02	$0.37
1/8/06	$0.39

4. What was the postal rate on July 4, 1994?
   F $0.29
   G $0.30
   H $0.31
   I $0.32

5. Which was probably the postal rate on January 1, 1985?
   A $0.10
   B $0.20
   C $0.25
   D $0.30

6. When the postal rate changes again, which is most likely to happen?
   F an increase of 2 or 3 cents
   G a decrease of 2 or 3 cents
   H an increase of 7 or 8 cents
   I an increase of 10 or 12 cents

NAEP Objective Test 8.4.1.a continued

Use this graph for questions 7–9.

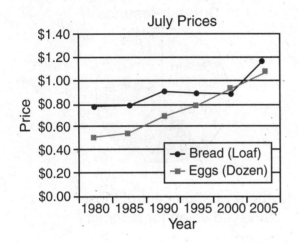

The bar graph compares some tall buildings. Use it for questions 10–12.

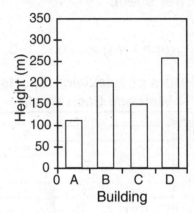

7. In what year was there the greatest difference in the price of the two items?
   A 1980
   B 1985
   C 1990
   D 1995

8. During what time period did the price of eggs increase the most?
   F 1980 to 1985
   G 1985 to 1990
   H 1990 to 1995
   I 2000 to 2005

9. Which was probably closest to the price of bread in July of 1993?
   A $0.65
   B $0.75
   C $0.85
   D $0.95

10. How many buildings are represented on the bar graph?
    F 4
    G 5
    H 50
    I 250

11. How many of the buildings are less than 250 meters tall?
    A 4
    B 3
    C 2
    D 1

12. From the bar graph, put the buildings in order from tallest to shortest.
    F D, B, A, C
    G A, C, B, D
    H D, B, C, A
    I C, A, B, D

NAEP Objective Test 8.4.1.c

Select the best answer for questions 1–12. Fill in the correct bubble on your answer sheet.

Use this graph for questions 1–3.

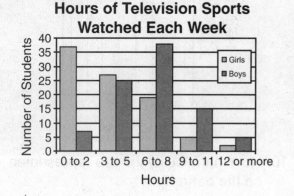

1. What is the approximate ratio of girls to boys who watched 0 to 2 hours of television sports?

   A 10 to 1
   B 5 to 1
   C 1 to 10
   D 1 to 5

2. About how many students watch fewer than 6 hours of sports each week?

   F 50   H 100
   G 75   I 125

3. Which is the best estimate of the total number of hours spent watching sports by all of the boys in the 6 to 8 hour category?

   A 100 to 125
   B 225 to 300
   C 325 to 500
   D 450 to 600

Use the circle graph for questions 4–6.

Survey Results

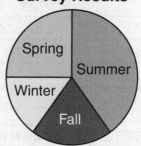

4. About what percent of those surveyed voted for summer?

   F 20%
   G 40%
   H 50%
   I 60%

5. Twenty people chose fall as their favorite season. About how many people were surveyed?

   A 50
   B 100
   C 150
   D 200

6. From this graph, what is the best estimate of the number of people who would choose spring as their favorite season if 500 people were surveyed?

   F 50
   G 125
   H 250
   I 325

Name _____ Date _____ Class _____

NAEP Objective Test 8.4.1.c continued

Use this graph for questions 7–9.

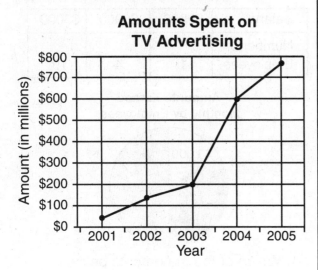

Use the graph for questions 10–12.

Amy and Todd each owed $300 on their new bicycles. The graph shows how they paid.

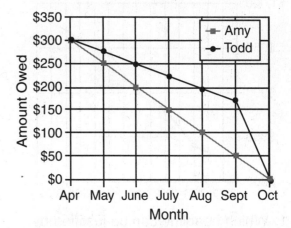

7. By what percent did the amount spent on television advertising increase from 2002 to 2003?
   A 25%
   B 50%
   C 100%
   D 150%

8. If the total advertising budget for 2004 was $950 million, approximately what fraction of the budget was spent on television advertising?
   F $\frac{1}{3}$
   G $\frac{1}{2}$
   H $\frac{2}{3}$
   I $\frac{3}{4}$

9. What is the best estimate of the total amount spent on television advertising over the five year period?
   A $1,200,000
   B $1,800,000
   C $950,000,000
   D $1,800,000,000

10. What is the difference between the monthly amounts Amy and Todd paid for the first five months?
    F $25
    G $50
    H $100
    I $125

11. How much more was Todd's final payment than Amy's?
    A $25
    B $50
    C $100
    D $125

12. In what month did Amy owe half as much as Todd?
    F May
    G June
    H August
    I September

Holt Mathematics Grade 6

Name _____ Date _____ Class _____

NAEP Objective Test 8.4.1.d

Select the best answer for questions 1–12. Fill in the correct bubble on your answer sheet.

1. Which headline can be justified by this graph?
   A Profits Increase Dramatically
   B Company Profits Sluggish
   C Slowing Trend in Company Profits
   D Company Profits Decline

2. Which of these graph changes would make the change in profits appear more prounounced?
   F Start the vertical scale at zero.
   G Bring the lines on the horizontal scale closer together.
   H Bring the lines in the vertical scale closer together.
   I Label the scales.

3. Which change in scale would make the company's profits appear flatter?
   A show only 2003 to 2004
   B show only 2001 to 2004
   C make scale $825 to $975
   D make scale $0 to $1200

This table shows employee salaries.

Salary	$9000	$6000	$3000
Number of Employees	1	4	10

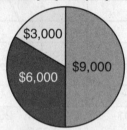

4. Which of the statements below can be justified by the graph and by the data?
   F Half of the employees earn $9000.
   G Most of the employees earn over $5000 per month.
   H Fewer than half of the employees earn $6000 per month.
   I $9000 is the most common employee salary.

5. What does the graph show?
   A The relative size of $3,000, $6,000, and $9,000
   B The percent of the employees who earn each amount
   C The percent of the total payroll earned by each employee
   D The comparison between number of employees and salary

6. In a graph showing the fraction of the employees who earn each salary, what is the fraction for $9000?
   F $\frac{1}{15}$ H $\frac{1}{9}$
   G $\frac{1}{14}$ I $\frac{1}{2}$

NAEP Objective Test 8.4.1.d continued

One Arbor Day, Glenn planted a one-year old tree that was 12 inches tall. Since then, he measures the tree each Arbor Day.

Age	Height
1 year	12 inches
2 years	25 inches
3 years	43 inches
4 years	63 inches
5 years	80 inches
6 years	98 inches

7. Which would be the most appropriate graph to show the data?
   A circle graph   C scatterplot
   B line graph   D line plot

8. Which would be the most suitable scale for a graph of the data?

   F 100, 80, 60, 40, 20, 0
   H 250, 200, 150, 100, 50, 0
   G 160, 110, 70, 40, 20, 10
   I 98, 80, 63, 43, 25, 12

9. Which title would most effectively describe the data in the graph?
   A Tree Growth   C Tree Age
   B Tree Height   D Arbor Day

Use this line graph for questions 10–12.

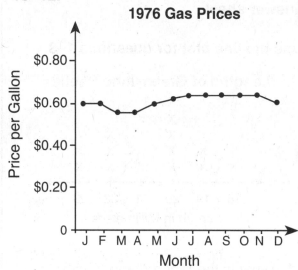

10. In which month was the price of gas highest?
    F August   H October
    G September   I Can not tell

11. How can the graph be changed to make the difference in the price of the gas from month to month more apparent?
    A Use a scale from $0.50 to $0.70.
    B Spell out the names of the months.
    C Make the points farther apart.
    D Change the title of the graph.

12. Which statement can be verified by this graph?
    F For two months in 1976, gas cost less than $0.60.
    G The average cost of gas in 1976 was $0.62.
    H Gas prices fluctuated drastically in 1976.
    I Gas prices stayed below $1 throughout the 1970s.

NAEP Objective Test 8.4.2.a

Select the best answer for questions 1–12. Fill in the correct bubble on your answer sheet.

Use the line plot for questions 1–3.

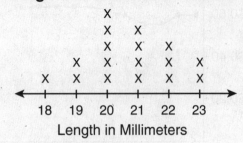

Lengths of Green June Beetles
Length in Millimeters

1. What is the range of the lengths of the green June beetles?
   A 5 mm
   B 6 mm
   C 17 mm
   D 21 mm

2. What is the mode of the green June beetle lengths?
   F 20 mm
   G 20.7 mm
   H 21 mm
   I 22 mm

3. What is the median green June beetle length?
   A 20 mm
   B 20.7 mm
   C 21 mm
   D 22 mm

4. Lucy said that more than half of the students in her class sleep at least 9 hours each night. Which measure of central tendency was she using?
   F mean
   G median
   H mode
   I range

5. Brad bought four shirts. The range of the prices was $7. The median price was $24.50. Which might be the prices of the four shirts?
   A $18, $24, $25, $22
   B $23, $26, $21, $28
   C $24, $24, $18, $26
   D $26, $32, $29, $26

6. All of Amy's math scores were between 76% and 97%. The mean score was 88%. Which were her scores?
   F 88%, 76%, 90%, 97%, 95%
   G 98%, 72%, 97%, 97%, 76%
   H 76%, 97%, 87%, 95%, 85%
   I 98%, 93%, 80%, 93%, 76%

NAEP Objective Test 8.4.2.a continued

The stem-and-leaf plot shows the time it took 20 people to complete a puzzle. Use it for questions 7–9.

```
3 | 1 2 4
2 | 0 0 1 3 6 6 8 8
1 | 2 3 4 8 8 8 9
0 | 9 9
```

7. What is the range of the times taken to complete the puzzle?
   A 23 minutes
   B 25 minutes
   C 28 minutes
   D 34 minutes

8. What is the mode of the times?
   F 18 minutes
   G 20 minutes
   H 28 minutes
   I 34 minutes

9. What is the median time taken to complete the puzzle?
   A 18 minutes
   B 20 minutes
   C 27 minutes
   D 28 minutes

The graph shows the height in inches of Kevin's high jumps.

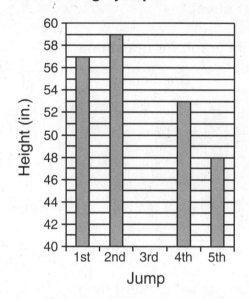

10. If the median height is 53 inches, which is a possible measurement for the third jump?
    F 50 in.
    G 54 in.
    H 57 in.
    I 60 in.

11. If the mean of the jump heights is 54 inches, what height was the third jump?
    A 50 in.   C 55 in.
    B 53 in.   D 57 in.

12. If the mode of the height measurements is 57 inches, what is the height of the third jump?
    F 48 in.
    G 53 in.
    H 57 in.
    I 59 in.

NAEP Objective Test 8.4.2.d

Select the best answer for questions 1–12. Fill in the correct bubble on your answer sheet.

Use this graph for questions 1–3.

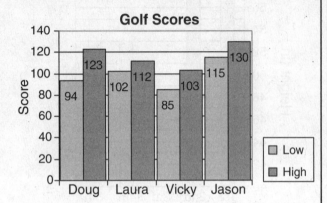

1. Who has the greatest range of golf scores?
   A Doug
   B Laura
   C Vicky
   D Jason

2. How many points greater is the mean of the high scores than the mean of the low scores?
   F 7 points
   G 18 points
   H 22 points
   I 45 points

3. What is the difference between the mean of Laura's two scores and the mean of Vicky's two scores?
   A 17 points
   B 11 points
   C 13 points
   D 9 points

Use this stem-and-leaf plot for questions 4–6.

Mole Masses in Grams

Starnose Mole		Hairytail Mole
3	7	
8 5 2 0	6	0 3 4
8 8 7 7 4	5	0 2 4 5 5 8 9
9 8 6 2	4	2 2 5 7 8
9 7	3	

4. How much greater is the range of the Starnose masses than the range of the Hairytail masses?
   F 2 g
   G 14 g
   H 16 g
   I 22 g

5. Which statement correctly compares the modes?
   A The mode for the Hairytail Mole is greater.
   B The mode is the same for the two types of mole.
   C The modes can not be compared because there is no mode for the Starnose Mole.
   D The mode for the Starnose Mole is greater.

6. What is the difference between the median masses of the two types of moles?
   F 2 g
   G 3 g
   H 4 g
   I 6 g

NAEP Objective Test 8.4.2.d continued

Use the double line graph for questions 7–12.

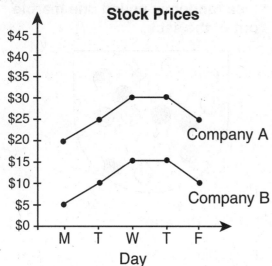

7. What is the difference in the price of Company A's and Company B's stock on Wednesday?
   A $45
   B $30
   C $15
   D $3

8. What is the mean price of Company A's stock for the five days?
   F $55
   G $26
   H $15
   I $5

9. What is the median of Company B's stock for the five days?
   A $10
   B $15
   C $20
   D $30

10. The range for Company A's stock is
    F 10 − 5 = 5
    G 15 − 5 = 10
    H 15 − 10 = 5
    I 30 − 20 = 10

11. The mean for Company B's stock is found by:
    A adding the values for the 5 days and dividing by 5
    B multiplying the range by 5
    C finding the most common value
    D finding the value in the middle

12. The values for Company A's and Company B's stock are always $15 apart.
    F The means are $15.
    G The means are the same.
    H The means are $15 apart.
    I The means are $30 apart.

NAEP Objective Test 8.4.4.b

Select the best answer for questions 1–12. Fill in the correct bubble on your answer sheet.

Use this information for questions 1–3.

Carl randomly pulls one block from the box.

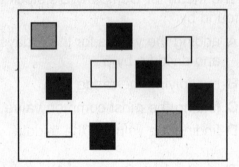

1. What is the probability that the block will be black?

   A $\dfrac{2}{5}$ C $\dfrac{2}{3}$

   B $\dfrac{1}{2}$ D $\dfrac{4}{5}$

2. What is the probability that the block will be black or white?

   F $\dfrac{3}{10}$

   G $\dfrac{3}{5}$

   H $\dfrac{2}{3}$

   I $\dfrac{4}{5}$

3. What is the probability that the block will be red?

   A 0   C $\dfrac{1}{5}$

   B $\dfrac{1}{10}$ D 1

Use this information for questions 4–6.

Lisa randomly pulled one marble out of the sack.

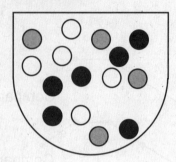

4. What is P(black)?

   F $\dfrac{1}{6}$

   G $\dfrac{1}{3}$

   H $\dfrac{2}{5}$

   I $\dfrac{2}{3}$

5. What is P(not white)?

   A $\dfrac{4}{15}$

   B $\dfrac{1}{3}$

   C $\dfrac{1}{2}$

   D $\dfrac{2}{3}$

6. What is P(solid color)?

   F 0

   G $\dfrac{1}{15}$

   H $\dfrac{2}{3}$

   I 1

NAEP Objective Test 8.4.4.b continued

Use this information for questions 7–12.

A number cube labeled 1–6 is tossed once, and this spinner is spun once.

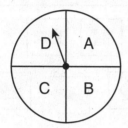

7. What is the probability of 6 and B?

   A $\frac{1}{24}$

   B $\frac{1}{23}$

   C $\frac{1}{6}$

   D $\frac{1}{4}$

8. What is the probability of an even number and D?

   F $\frac{1}{24}$

   G $\frac{1}{8}$

   H $\frac{1}{2}$

   I $\frac{3}{4}$

9. P(4 and E) = ?

   A 0

   B $\frac{1}{30}$

   C $\frac{1}{24}$

   D $\frac{1}{6}$

10. What is the probability of 3 and a consonant?

    F $\frac{1}{24}$

    G $\frac{1}{8}$

    H $\frac{3}{8}$

    I $\frac{6}{3}$

11. What is P(2 and not C)?

    A $\frac{1}{24}$

    B $\frac{1}{8}$

    C $\frac{1}{3}$

    D $\frac{2}{3}$

12. What is the probability of a number greater than 2 and A or B?

    F $\frac{1}{6}$

    G $\frac{1}{4}$

    H $\frac{1}{3}$

    I $\frac{2}{3}$

NAEP Objective Test 8.4.4.e

Select the best answer for questions 1–12. Fill in the correct bubble on your answer sheet.

1. In an experiment this spinner is spun once. Which shows the sample space for this experiment?

   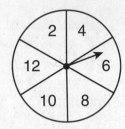

   A the even numbers less than 13
   B 1, 2, 3, 4, 5, 6, 7, 8, 9, 10, 11, 12
   C 2, 4, 6, 8, 10, 12
   D 6

2. Michael wrote each letter of his name on a card and put the cards in a bag. He will pull one card out of the bag at random. What is the sample space for this experiment?

   F 7
   G Michael
   H consonants, vowels
   I M, I, C, H, A, E, L

3. A bag contains 2 orange, 3 yellow, and 2 blue marbles. One marble will be pulled out of the bag at random. Which outcome is **not** part of the sample space for this experiment?

   A yellow
   B blue and yellow
   C blue
   D orange

4. Javier will pull one shape card out of the box at random. How many different outcomes are in the sample space for this experiment?

   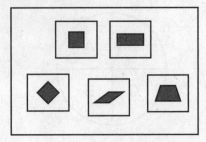

   F 1
   G 2
   H 4
   I 5

5. Brittany wrote each of the first four odd numbers in a section of a 4-part spinner. She will spin the spinner once. Which outcome is **not** part of the sample space for this experiment?

   A 1   C 4
   B 3   D 5

6. This spinner is spun one time. How many different outcomes are in the sample space for this experiment?

   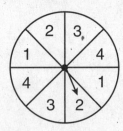

   F 1   H 4
   G 2   I 8

NAEP Objective Test 8.4.4.e continued

7. A penny is tossed and this spinner is spun once. Which shows the sample space for this experiment?

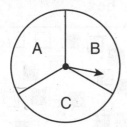

   A AB, AC, BC
   B heads, A; heads, B; heads, C; tails, A; tails, B; tails, C
   C heads, letter; tails, letter
   D penny, A; penny, B; penny, C

8. Two number cubes each labeled 1–6 are tossed at the same time. Which outcome is in the sample space for this experiment?
   F 2, 3
   G 4, 7
   H 10
   I 36

9. A penny and a nickel are tossed at the same time. Which outcome is **not** in the sample space for this experiment?
   A heads, tails
   B penny, nickel
   C tails, heads
   D tails, tails

10. Ryan has a pair of black socks, a pair of blue socks, and a pair of brown socks in a drawer. He pulls 2 socks out of the drawer at one time. How many different outcomes are possible?
    F 12
    G 9
    H 6
    I 3

11. Chloe wrote each letter of her name on a card and put the cards in a bag. She will pull one card out of the bag at random, put it back, and pull out another card. How many different outcomes are in the sample space for this experiment?
    A 2   C 10
    B 5   D 25

12. Ben will pull out one card at random from each box. How many different outcomes are in the sample space for this experiment?

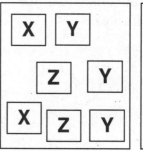 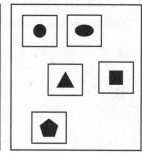

    F 35
    G 15
    H 12
    I 7

NAEP Objective Test 8.4.4.g

Select the best answer for questions 1–12. Fill in the correct bubble on your answer sheet.

1. A weather forecaster stated that the chance of rain today is 90%. What is the probability that it will rain today?

   A $\dfrac{9}{100}$

   B $\dfrac{1}{9}$

   C $\dfrac{9}{10}$

   D $\dfrac{9}{1}$

2. Of all the students in Ms. Byers' class, 0.44 are bilingual. If Ms. Byers chooses one of the students at random, what is the probability that she will choose a bilingual student?

   F 4%
   G 44%
   H 50%
   I 56%

3. Which of these fractions could **not** represent a probability?

   A $\dfrac{4}{2}$

   B $\dfrac{3}{3}$

   C $\dfrac{1}{5}$

   D $\dfrac{0}{7}$

Use this information for questions 4–6.

Miguel randomly pulls one block from the box.

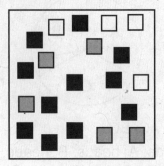

4. What is the probability that the block will be grey?

   F 5%

   G $\dfrac{1}{4}$

   H $\dfrac{1}{3}$

   I 0.5

5. What is the probability that the block will be black or white?

   A $\dfrac{2}{3}$

   B 75%

   C 3.4

   D 4.11

6. What is the probability that the block will **not** be black?

   F 9%

   G 0.45

   H $\dfrac{9}{11}$

   I 90%

NAEP Objective Test 8.4.4.g continued

7. Which of these decimals could not represent a probability?
   A 0.1
   B 0.14
   C $0.\overline{3}$
   D 1.4

8. Amy placed these cards in a bag. She pulled out one card at random. Which does not represent the probability that she pulled out an E?

   F 3%
   G 0.3
   H 30%
   I $\frac{3}{10}$

9. A spinner is spun once. The probability of spinning a 3 is $\frac{1}{2}$. Which statement is true?
   A P(3) = 0.2
   B P(3) = 0.3
   C P(3) = 0.5
   D P(3) = 1.2

10. Which of these percents could not represent a probability?
    F 0%
    G 3%
    H 30%
    I 300%

11. The name of each student at Keller Middle School is placed in a bag. The name of one student is chosen at random. What is the probability that a sixth grader will be chosen?

    Keller Middle School

    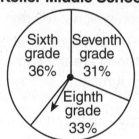

    A $0.\overline{3}$
    B $\frac{1}{3}$
    C 0.36
    D $\frac{9}{16}$

12. The five sections of a spinner are labeled 1, 1, 2, 3, and 3. The spinner is spun one time. Which does **not** represent the probability of an odd number?
    F 80%
    G 0.8
    H $\frac{4}{5}$
    I 0.4

NAEP Objective Test 8.4.4.j

Select the best answer for questions 1–12. Fill in the correct bubble on your answer sheet.

Use this information to answer questions 1–3.

> TODAY'S WEATHER
> 60% chance of rain

1. Which of these statements is true?
   A It will rain for 60% of today.
   B There will be 0.60 inch of rain today.
   C Today is the 60th day of the year on which it will rain.
   D On 60% of the days when conditions are the same as today, it rains.

2. How likely is it that it will rain today?
   F It is unlikely that it will rain.
   G It will not rain.
   H It is more likely to rain than not rain.
   I It is certain that it will rain.

3. What is the probability that it will not rain?
   A 40%
   B 50%
   C 60%
   D 100%

Use this information for questions 4–6.

A bag contains 10 marbles. Kari is finding the probability of pulling each different color marble out of the bag at random. So far, she has recorded these probabilities.

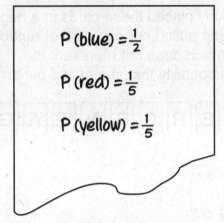

$P(\text{blue}) = \frac{1}{2}$

$P(\text{red}) = \frac{1}{5}$

$P(\text{yellow}) = \frac{1}{5}$

4. How many marbles are blue?
   F 1
   G 2
   H 5
   I 8

5. How many marbles are red?
   A 1
   B 2
   C 5
   D 6

6. How many different colors of marbles are in the bag?
   F 3
   G 4
   H 5
   I 10

NAEP Objective Test 8.4.4.j continued

Use this information for questions 7–9.

A box contains different colored blocks. One block is pulled out at random. The probability that the block will be yellow is $\frac{5}{12}$.

7. What is the fewest number of blocks that could be in the box?
   A 5
   B 11
   C 12
   D 24

8. Which of these could **not** be the number of yellow blocks in the box?
   F 5
   G 10
   H 12
   I 15

9. Suppose there are 48 blocks in the box. How many of the blocks are yellow?
   A 5
   B 9
   C 15
   D 20

Use this information for questions 10–12.

A number cube is labeled 1–6, and this spinner has a letter in each section. The cube is tossed once, and the spinner is spun once.

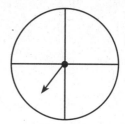

10. If $P(3, B) = \frac{1}{24}$, how many sections of the spinner are labeled B?
    F 1
    G 2
    H 3
    I 4

11. If $P(4, A) = \frac{1}{8}$, how many sections of the spinner are labeled A?
    A 1
    B 2
    C 3
    D 4

12. Which of these probabilities is greatest?
    F P(even number, B)
    G P(number less than 5, B)
    H P(multiple of 6, B)
    I P(7, B)

NAEP Objective Test 8.5.1.a

Select the best answer for questions 1–12. Fill in the correct bubble on your answer sheet.

1. Which two numbers come next in the pattern?

   0, 3, 7, 12, 18, …

   A 25, 33
   B 25, 32
   C 28, 33
   D 27, 34

2. Keith has a bank account. Each week he deposits more than he did the week before. His deposits are listed below. Following this pattern, how much will he deposit in Week 5?

Week 1	Week 2	Week 3	Week 4	Week 5
$15.00	$28.00	$43.00	$60.00	?

   F $68.00
   G $72.00
   H $79.00
   I $83.00

3. Which two shapes come next in this pattern?

   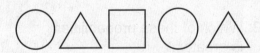

   A ○ △
   B △ □
   C ○ □
   D □ ○

4. Which is the missing number in the sequence?

   32, 44, 56, ____, 80, 92

   F 64
   G 68
   H 72
   I 74

5. The table shows the relationship between the weight of a bag of fruit in pounds and its weight in ounces.

   Fruit Weights

Weight (lb)	2	3	4	5	6
Weight (oz)	32	48	64	80	?

   What will a 6-pound bag of fruit weigh?

   A 60 ounces
   B 84 ounces
   C 92 ounces
   D 96 ounces

6. Which two numbers come next in the pattern?

   27, 23, 19, 15, 11, …

   F 9, 5
   G 7, 3
   H 6, 3
   I 8, 4

NAEP Objective Test 8.5.1.a continued

7. The table shows the perimeters of rectangles where the lengths stay the same but the widths change.

Length (in.)	Width (in.)	Perimeter (in.)
5	2	14
5	3	16
5	4	18
5	5	20
5	6	?

What will the perimeter of a 5-inch by 6-inch rectangle be?

A 20 inches
B 22 inches
C 24 inches
D 30 inches

8. Which shape comes next in this pattern?

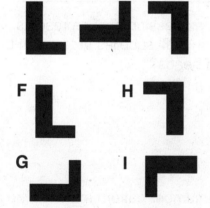

9. Which is the missing number in the sequence?

2, 8, 18, ____, 50, 72

A 28
B 32
C 40
D 46

10. Pauline is training for a race. On April 1, she runs for 15 minutes. Every day she runs for 2 more minutes than she did on the previous day. How many minutes will Pauline run on April 5?

Day	Time (min)
April 1	15
April 2	17
April 3	19
April 4	
April 5	?

F 23
G 25
H 27
I 29

11. Which number comes next in the pattern?

2.2, 2.4, 2.6, 2.8, ?

A 2.0
B 2.6
C 3.0
D 3.2

12. Jin is saving money for a new baseball glove. He puts $5.00 each week aside. How much money will Jin have put aside after 4 weeks?

F $15.00
G $20.00
H $25.00
I $40.00

Name _____ Date _____ Class _____

NAEP Objective Test 8.5.1.c

Select the best answer for questions 1–12. Fill in the correct bubble on your answer sheet.

1. Identify the best rule for this pattern.

   | 4 | 8 | 16 | 32 | 64 |

   A The numbers are increasing by 4.
   B The numbers are increasing by 8.
   C The numbers are doubling.
   D The numbers are tripling.

2. Use the following pattern to find the units digit of 4^8.

   $4^1 = \underline{4}$
   $4^2 = 1\underline{6}$
   $4^3 = 6\underline{4}$
   $4^4 = 25\underline{6}$
   $4^5 = 1{,}02\underline{4}$

   F 2
   G 4
   H 6
   I 8

3. Which expression best represents the y values in terms of the x values?

x	1	2	3	4	5	6
y	3	5	7	9	11	13

   A $x + 2$
   B $x + 7$
   C $\dfrac{(x-1)}{2}$
   D $2x + 1$

4. Which expression is the rule for the data, d, in the table?

h	d
3	16
5	26
7	36

   F $h + 13$
   G $5h + 1$
   H $4h + 6$
   I $6h - 3$

5. The table shows the amounts Dana spent for beads to make four necklaces. It also shows the amounts she spent for all the supplies to make each necklace.

   Necklace Costs

Beads, x ($)	6	10	15	22
All supplies, y ($)	12	16	21	28

   Which expression best represents the cost of all supplies in terms of the cost of beads?

   A $x + 6$
   B $y - 6$
   C $2x$
   D $y - 2$

6. Which number pattern matches the following rule?

   The numbers are increasing by 3.

   F 0, 3, 5, 8
   G 4, 7, 10, 13
   H 6, 9, 15, 18
   I 18, 15, 12, 9

NAEP Objective Test 8.5.1.c continued

7. The table shows the amounts Joe earned caring for neighbors' pets over four weeks. It also shows the amount he had left after putting some money in savings.

   Ed's Earnings

Ed's earnings, x ($)	20	24	30	36
Amount left, y ($)	10	14	20	26

   Which expression best represents the amount left in terms of the amount earned?

   A $y + 10$
   B $x - 10$
   C $2y$
   D $\dfrac{x}{2}$

8. Which number pattern matches the following rule?

   The numbers are decreasing by 4.

   F 20, 16, 12, 8
   G 23, 19, 16, 12
   H 30, 24, 18, 12
   I 32, 28, 16, 8

9. Which expression is the rule for the data, d, in the table?

r	d
2	5
4	9
6	13

   A $r + 3$
   B $2r + 1$
   C $3r - 1$
   D $r^2 + 1$

10. Which expression best represents the y values in terms of the x values?

x	1	2	3	4	5	6
y	2	6	10	14	18	22

    F $x + 4$
    G $2x$
    H $4x - 2$
    I $\dfrac{(x + 2)}{4}$

11. Identify the best rule for this pattern.

1	3	9	27	81

    A The numbers are increasing by 2.
    B The numbers are increasing by 6.
    C The numbers are doubling.
    D The numbers are tripling.

12. A particular recipe calls for 2 eggs, 1 cup of flour, and 2 teaspoons of oil. Which of the following amounts would you use if you wanted to double the recipe?

    F 1 egg, $\dfrac{1}{2}$ c flour, 1 tsp oil
    G 2 eggs, 2 c flour, 2 tsp oil
    H 4 eggs, 1 c flour, 2 tsp oil
    I 4 eggs, 2 c flour, 4 tsp oil

NAEP Objective Test 8.5.1.e

Select the best answer for questions 1–12. Fill in the correct bubble on your answer sheet.

1. In the equation $y = 4x$ if the value of x is increased by 1, what is the effect on the value of y?

   A It is 2 times the original amount.
   B It is 4 times the original amount.
   C It is 2 more than the original amount.
   D It is $\frac{1}{4}$ of the original amount.

2. Jared has a bank account. Every month he deposits the same amount of money. His balances so far for this year are shown below. Which of the following statements is true?

Jan.	Feb.	Mar.	Apr.
$25.00	$37.50	$50.00	$62.50

   F Jared deposits $12.50 every month.
   G Jared deposits $17.50 every month.
   H Jared deposits $25.00 every month.
   I Jared will have $80.00 in his account in May.

3. In the equation $y = x + 5$ if the value of x is increased by 2, what is the effect on the value of y?

   A It is 2 times the original amount.
   B It is 2 more than the original amount.
   C It is 5 more than the original amount.
   D It is 7 more than the original amount.

4. In the graph below, if the value of x is increased by 1, what is the effect on the value of y?

   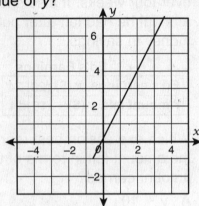

   F y is increased by 1.
   G y is increased by 2.
   H y is increased by 4.
   I y is increased by 6.

5. A pancake recipe calls for different amounts of mix and water depending on the number of servings needed. The table shows the amounts of mix and water per serving.

Servings	5	10	15	20	25
Mix (cups)	1	2	3	4	5
Water (cups)	$\frac{1}{2}$	1	$1\frac{1}{2}$	2	?

   At this rate, how many cups of water are needed for 25 servings?

   A $2\frac{1}{2}$ C $3\frac{1}{2}$
   B 3   D 5

6. If x represents the number of newspapers Kim delivers each day, which of the following represents the total number of newspapers Kim delivers in 4 days?

   F $4x$ G $x + 4$
   H $\frac{(x+x+x+x)}{4}$ I $x + 5$

NAEP Objective Test 8.5.1.e continued

7. The table below shows the *x*- and *y*-coordinates of some ordered pairs.

x	2	3	4	5
y	5	8	11	14

If the value of *x* is increased by 1, what is the effect on the value of *y*?

 A *y* is increased by 1.
 B *y* is increased by $2\frac{1}{2}$.
 C *y* is increased by 3.
 D *y* is increased by 5.

8. In the equation $y = x + 4$ if the value of *x* is increased by 1, what is the effect on the value of *y*?

 F It is 1 more than the original amount.
 G It is 2 more than the original amount.
 H It is 4 more than the original amount.
 I It is 5 more than the original amount.

9. In the graph below, if the value of *x* is increased by 1, what is the effect on the value of *y*?

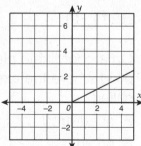

 A *y* is increased by $\frac{1}{2}$.
 B *y* is increased by 1.
 C *y* is increased by 2.
 D *y* is increased by 4.

10. Ingrid records the height of her plant each week. If the plant continues to grow at the same rate, what will the difference in height be between the fourth and fifth week?

Week	Plant Height (cm)
1	3
2	5
3	7
4	9
5	?

 F 1 cm
 G 2 cm
 H 9 cm
 I 11 cm

11. Nadine has an exercise program where each day she does 20 sit-ups plus an additional number equal to the day of the month. During the month, how does the number of sit-ups change from one day to the next day?

 A The number increases by 1.
 B The number increases by 2.
 C The number increases by 20.
 D The number increases by the number of days in that month.

12. In the equation $y = 3x$ if the value of *x* is increased by 2, what is the effect on the value of *y*?

 F It is 2 times the original amount.
 G It is 3 more than the original amount.
 H It is 5 times the original amount.
 I It is 6 more than the original amount.

Name _____ Date _____ Class _____

NAEP Objective Test 8.5.2.c

Select the best answer for questions 1–12. Fill in the correct bubble on your answer sheet.

1. Find the coordinates of point J.

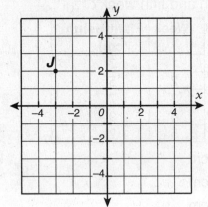

A (2, −3)   C (3, −2)
B (−2, −3) D (−3, 2)

2. Which point is plotted at (−3, −3) on the grid?

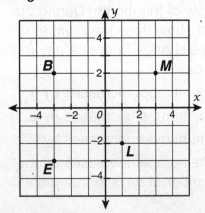

F point B   H point M
G point E   I point L

3. What is the x-coordinate of the ordered pair (−4, 3)?
A −4   C 3
B −3   D 4

4. Which point is plotted at (1, −2) on the grid?

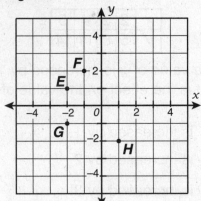

F point E   H point G
G point F   I point H

5. Find the coordinates of point K.

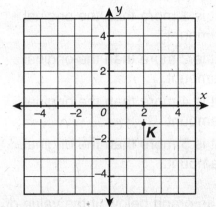

A (2, −1)   C (−1, −2)
B (1, −2)   D (−2, 1)

6. On a coordinate plane, what is the point where the axes intersect called?
F quadrant
G origin
H ordered pair
I coordinate point

NAEP Objective Test 8.5.2.c continued

7. What is the y-coordinate of the ordered pair (2, −5)?
   A −5
   B −2
   C 2
   D 5

8. Which point is plotted at (3, 2) on the grid?

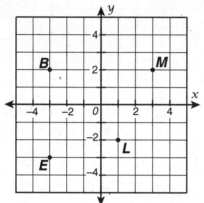

   F point B
   G point E
   H point M
   I point L

9. Find the coordinates of point L.

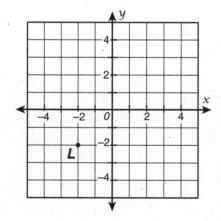

   A (2, 2)
   B (−2, 2)
   C (2, −2)
   D (−2, −2)

10. Which point is plotted at (−2, 1) on the grid?

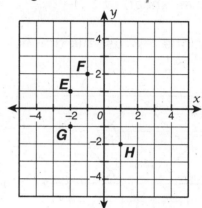

    F point E
    G point F
    H point G
    I point H

11. Find the coordinates of point R.

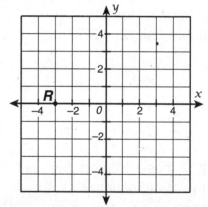

    A (0, 3)
    B (0, −3)
    C (3, 0)
    D (−3, 0)

12. What is the y-coordinate of a point plotted on the x axis?
    F x
    G y
    H 0
    I 1

NAEP Objective Test 8.5.2.d

Select the best answer for questions 1–12. Fill in the correct bubble on your answer sheet.

For questions 1–3, use the graph below.

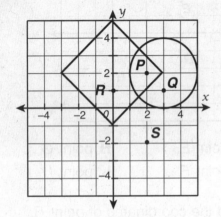

1. Which point is located at the coordinate pair (2, −2)?
   A P
   B Q
   C R
   D S

2. Which of these ordered pairs is NOT a vertex of the square?
   F (2, 2)   H (3, 2)
   G (0, −1)   I (−3, 2)

3. Which ordered pair locates a point inside both the square and the circle?
   A P
   B Q
   C R
   D S

For questions 4–6, use the graph below.

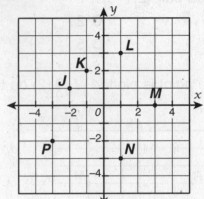

4. Which ordered pair represents the result of sliding point K three units down?
   F (−3, −1)
   G (−2, 1)
   H (−1, −1)
   I (2, 1)

5. Find the coordinates of point N.
   A (0, 3)
   B (1, −3)
   C (−3, 1)
   D (1, 3)

6. Which point is plotted at (−2, 1) on the grid?
   F J
   G K
   H N
   I P

NAEP Objective Test 8.5.2.d continued

For questions 7 and 8, use the graph below.

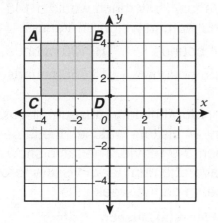

7. What is the coordinate of B' after a reflection of B across the y-axis?

   A (−1, 3)
   B (4, 4)
   C (1, 4)
   D (1, 1)

8. What is the coordinate of A' after a reflection across the x-axis?

   F (2, 4)
   G (4, −4)
   H (−4, 4)
   I (−4, −4)

9. When the point (−3, 2) is translated four units to the right and three units down, what are the new coordinates?

   A (4, −3)
   B (1, −1)
   C (−6, 6)
   D (−7, −1)

10. When the point (0, 3) is reflected across the x-axis, what are the new coordinates?

    F (3, 0)
    G (−3, 0)
    H (0, 3)
    I (0, −3)

For questions 11 and 12, use the graph below.

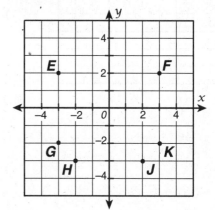

11. Which point represents a reflection of point G over the x-axis?

    A E
    B F
    C K
    D H

12. Which ordered pair represents a translation of point J two units to the left and three units up?

    F (4, 0)
    G (−1, −1)
    H (0, 0)
    I (−2, 3)

NAEP Objective Test 8.5.2.g

Select the best answer for questions 1–12. Fill in the correct bubble on your answer sheet.

1. Which equation can be used to describe the graph below?

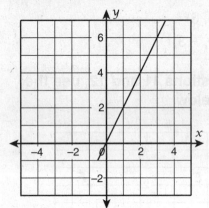

   A $y = x + 2$
   B $y = \dfrac{1}{2}x$
   C $y = 2x$
   D $y = 3x - 1$

2. Ariel deposits the same amount into her bank each week. The table below show her balance for the last five weeks.

Week 1	Week 2	Week 3	Week 4	Week 5
$12.00	$24.00	$36.00	$48.00	$60.00

   Which equation describes her balance in dollars, b, for any week, w?

   F $b = 12w$
   G $b = w + 12$
   H $\dfrac{w}{b} = 12$
   I $b = 60 - w$

3. Jaime's family drinks 3 gallons of milk every 4 days. In a month with 31 days, about how many gallons of milk will Jaime's family drink?

   A 23 gallons
   B 24 gallons
   C 26 gallons
   D 93 gallons

4. An object that weighs 10 pounds on Earth would weigh about 9 pounds on Venus. How much would a 140-pound woman weigh on Venus?

   F 14 pounds
   G 126 pounds
   H 130 pounds
   I 156 pounds

5. At a natural history museum, 1 out of every 14 visitors watches the large-screen 3-D movie. The table shows the average number of visitors to the museum on the weekends.

   Weekend Museum Visitors

Day	Number of visitors
Saturday	1,565
Sunday	1,842

   Which proportion can be used to find x, the number of visitors who can be expected to watch a large-screen 3-D movie on a Saturday?

   A $\dfrac{1}{x} = \dfrac{1,565}{14}$
   B $\dfrac{x}{1} = \dfrac{14}{1,565}$
   C $\dfrac{1}{14} = \dfrac{x}{1,565}$
   D $\dfrac{1}{14} = \dfrac{1,565}{x}$

6. The table below shows the x- and y-coordinates of some ordered pairs.

x	7	8	9	10
y	10	12	14	16

   Which equation describes the relationship of the x values to the y values?

   F $y = x + 2$
   G $y = x + 3$
   H $y = 2x - 4$
   I $y = 3x - 11$

NAEP Objective Test 8.5.2.g continued

7. Judy is making a drawing of her school using the following scale:

   1 inch = 25 feet

   The length of the school is 375 feet. How many inches long should Judy make the school in her drawing?

   A 15 inches
   B 15 feet
   C 25 inches
   D 50 inches

8. Marco bowls over 100 in 4 out of every 10 games. During a week's practice Marco bowled 25 games. Which proportion can be used to find x, the expected number of times Marco bowls over 100?

   F $\dfrac{4}{x} = \dfrac{10}{25}$ H $\dfrac{4}{10} = \dfrac{x}{25}$

   G $\dfrac{4}{10} = \dfrac{25}{x}$ I $\dfrac{4}{25} = \dfrac{x}{10}$

9. Which equation can be used to describe the graph below?

   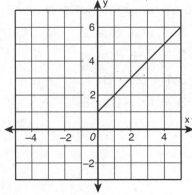

   A $y = x$
   B $y = x + 1$
   C $y = x - 1$
   D $y = 2x + 1$

10. At a 10-mile road race, there is a water station every $\dfrac{1}{2}$ mile and at the finish line. Which formula can be used to find x, the total number of water stations at the race?

    F $10x = \dfrac{1}{2}$

    G $\dfrac{1}{2}x = 10$

    H $x = \dfrac{1}{2} \cdot 10$

    I $x = 10 - \dfrac{1}{2}x$

11. Connie sells personalized placemats through the mail. She charges $8.50 per placemat plus $4.00 per order for shipping. Which expression describes the cost in dollars and cents of p placemats, including shipping?

    A 12.5p
    B 4p + 8.5
    C 8.5p + 4
    D 12.5p + 4

12. Which equation could you solve to find the value of x in the proportion $\dfrac{3}{4} = \dfrac{x}{32}$?

    F $3 \cdot 4 = 32x$
    G $3 \cdot 32 = 4x$
    H $3x = 4 \cdot 32$
    I $3x = 3 \cdot 32$

NAEP Objective Test 8.5.3.b

Select the best answer for questions 1–12. Fill in the correct bubble on your answer sheet.

1. Which expression describes the following sequence?

1	2	3	4	n
5	8	11	14	

   A $2n + 3$
   B $3n + 2$
   C $4n + 1$
   D $5n - 1$

2. Andy sees this lunch special in a restaurant window.

   Which expression gives the price for Andy and two friends if each gets the special and each has 3 meatballs?

   F $3 \cdot \$4.50 + 9 \cdot 0.75$
   G $9 \cdot (\$4.50 + 0.75)$
   H $3 \cdot \$4.50 + 0.75$
   I $3 \cdot 0.75 + \$4.50$

3. There are 5,280 feet in 1 mile. Which equation can be used to find x, the number of feet in 5 miles?

   A $x = 5{,}280 \cdot 5$
   B $x = 5{,}280 + 5$
   C $x = 5{,}280 - 5$
   D $x = 5{,}280 \div 5$

4. Quick Clean is a house cleaning service. The service charges $15 per room plus $12 per hour after the first hour. Which expression describes how to find the total cost?

   F $15r + 12h$
   G $15r + 12(h - 1)$
   H $15r + 12(h + 1)$
   I $15r(12h)$

5. Louis needs 25 more hours of training to become a lifeguard. He has already received 8 hours of training. Which equation shows the total number of training hours Louis needs to become a lifeguard?

   A $25 + x = 8$
   B $x - 8 = 25$
   C $8 - x = 25$
   D $25 - x = 8$

6. Anita owns three brown skirts, w white skirts, and two red skirts. Which equation shows how to find how many white skirts she owns if she owns eight skirts in all?

   F $3 + w + 8 = 2$
   G $3 + 2 + 8 = w$
   H $3w + 2 = 8$
   I $3 + w + 2 = 8$

NAEP Objective Test 8.5.3.b continued

7. The population in Carla's town is three times smaller than the population in Tonya's town. Which equation correctly represents the population if y represents the population in Carla's town and x represents the population in Tonya's town?

   A $y = \left(\frac{1}{3}\right)x$
   B $y = 3x^2$
   C $y = 3x$
   D $y = x - 3$

8. Ms. Ramirez is a floral designer. She is making 3 large arrangements that will have the same number of roses and lilies in each. She has 54 roses that she can use in the arrangements. Which equation can be used to find r, the number of lilies she can use in each arrangement?

   F $r = 54 \cdot 3$ H $r = 54 \div 3$
   G $r = 54 - 3$ I $r = 54 + 3$

9. The sign shows the cost of renting a canoe.

CANOE RENTALS	
Basic Fee	$9.00
Hourly rate	$4.50

   Which equation can be used to find c, the cost in dollars of renting the canoe for h hours?

   A $c = 4.5(9 + h)$
   B $c = 9h + 4.5$
   C $c = 9 + 4.5h$
   D $c = (9 + 4.5)h$

10. The price tag shows the regular price and the sale price of a pair of jeans. Mr. Williams bought a pair of these jeans on sale for each of his 3 daughters.

    Which equation can be used to find s, the total amount in dollars that Mr. Williams saved by buying 3 pairs of jeans on sale?

    F $s = 29.5 - 22.99$
    G $s = 3(29.5) - 22.99$
    H $s = 3(22.99) - 3(29.5)$
    I $s = 3(29.5) - 3(22.99)$

11. Look at the input-output table. What is the rule for this table?

x	y
4	9
6	13
8	17

    A $y = 2x + 1$
    B $y = -2x + 18$
    C $y = 3x - 5$
    D $y = x + 11$

12. There are 36 inches in 1 yard. Which equation can be used to find x, the number of yards in 288 inches?

    F $x = 288 \cdot 36$
    G $x = 288 - 36$
    H $x = 288 + 36$
    I $x = 288 \div 36$

NAEP Objective Test 8.5.3.c

Select the best answer for questions 1–12. Fill in the correct bubble on your answer sheet.

1. What is the value of 4^2?
   - A 2
   - B 4
   - C 8
   - D 16

2. The rule for the data, d, in the table below is $4h - 3$.

h	d
2	5
4	13
6	21
8	

   What is the data value for 8?
   - F 28
   - G 29
   - H 32
   - I 35

3. Use the order of operations to find which expression is true.
   - A $8 + 12(14 - 4) + 6 = 206$
   - B $10 - 8 \cdot 6 + 7 = 26$
   - C $14 + 6(7 - 3) - 8 = 30$
   - D $4 + 5(6 + 2) - 5 = 66$

4. What whole number when raised to the third power equals 64?
   - F 2
   - G 4
   - H 6
   - I 8

5. Neela buys blank CDs online that cost $8 for a package of 50. Shipping costs an additional $3 per order. Using the formula below, what will Neela pay for 3 packages of CDs?

   $8c + 3$

   - A $24
   - B $27
   - C $33
   - D $48

6. Use the order of operations to find the value of the expression.

   $58 + 8\left(\dfrac{45}{9}\right) - 21$

   - F 42
   - G 77
   - H 225
   - I 309

104

Holt Mathematics Grade 6

NAEP Objective Test 8.5.3.c continued

7. What is the value of the expression $15s - 4t$ when $s = 6$ and $t = 2$?
   A 16
   B 19
   C 72
   D 82

8. **MAY 2006**

Su	M	Tu	W	Th	F	Sa
	1	2	3	4	5	6
7	8	9	10	11	12	13
14	15	16	17	18	19	20
21	22	23	24	25	26	27
28	29	30	31			

   You can convert from weeks, w, to days, d, by multiplying by 7 as in the equation below.

   $7w = d$

   How many days are there in 26 weeks?
   F $\frac{7}{26}$
   G $\frac{26}{7}$
   H 26
   I 182

9. What is the value of $m \cdot (-3)$ when $m = -7$?
   A -21
   B -10
   C 4
   D 21

10. Use the order of operations to find the value of the expression.

    $20 + 25 \div 5 - 4 \cdot 2$

    F 1
    G 10
    H 17
    I 22.5

11. In the equation $y = -7x + 3$, what is the value of y when $x = -1$?
    A -4
    B 3
    C 10
    D 11

12. What is the value of $3^2 + 4$?
    F 7
    G 10
    H 11
    I 13

NAEP Objective Test 8.5.4.a

Select the best answer for questions 1–12. Fill in the correct bubble on your answer sheet.

1. Which of the following equations has a solution of 6?

   A $x + 13 = 26$
   B $x + 24 = 32$
   C $x + 42 = 50$
   D $x + 15 = 21$

2. Ursula is at the grocery store and sees the following:

   Oranges: 50 cents each
   One dozen for $3.60

   Ursula wants to know the price, x, per orange if she buys a dozen. She uses the following equation.

   $12x = \$3.60$

   What is the value of x?

   F 30 cents
   G 36 cents
   H 42 cents
   I 48 cents

3. Which of the following is the solution to the equation $15x = 225$?

   A $x = 10$
   B $x = 12$
   C $x = 15$
   D $x = 18$

4. What is the value of d in the following equation?

   $\dfrac{d}{4} = 12$

   F $d = 3$
   G $d = 16$
   H $d = 24$
   I $d = 48$

5. Which of the following equations does NOT have a solution of $x = 4$?

   A $13x = 52$
   B $x - 6 = 2$
   C $x + 16 = 20$
   D $\dfrac{x}{2} = 2$

6. Jessie wants to solve the equation $4x = 112$. What step should she take first?

   F Add 4 to both sides.
   G Subtract 4 from both sides.
   H Multiply 4 to both sides.
   I Divide by 4 on both sides.

NAEP Objective Test 8.5.4.a continued

7. Which of the following is a solution to the equation $x + 6 = 24$?

A $x = 4$
B $x = 18$
C $x = 30$
D $x = 36$

8. A mechanic shop posts the following sign that gives the cost for auto repairs.

This can be written as the following formula.

Cost = $85 + 55h$, where h is the number of hours that the repair takes.

If the cost of a repair was $525, how many hours did the repair take?

F 6 hours
G 7 hours
H 8 hours
I 9 hours

9. Sam has six blue shirts, w white shirts, and four red shirts. How many white shirts does he have if he has 15 shirts in all?

A 5 shirts
B 6 shirts
C 10 shirts
D 11 shirts

10. Which value of y makes the equation true?

$45.092 + y = 78.1$

F $y = 33.008$
G $y = 33.8$
H $y = 122.12$
I $y = 123.19$

11. What is the solution to the following equation?

$x - 7 = -16$

A $x = -23$
B $x = -9$
C $x = 9$
D $x = -112$

12. Which of the following equations does NOT have a solution of $x = 6$?

F $2x - 4 = 10$
G $\left(\frac{1}{3}\right)x \cdot 8 = 16$
H $3x - 5 = 13$
I $x^2 = 36$

NAEP Objective Test 8.5.4.c

Select the best answer for questions 1–12. Fill in the correct bubble on your answer sheet.

1. What is the solution to the following equation?

   $\left(\frac{3}{4}\right)x - 5 = 79$

   A $x = 63$
   B $x = 108$
   C $x = 112$
   D $x = 116$

2. Which of the following equations does NOT have a solution of $x = 9$?

   F $\frac{2}{3}(x) - 4 = 10$
   G $\left(\frac{1}{3}\right)x \cdot 8 = 24$
   H $3x - 5 = 22$
   I $\left(\frac{1}{3}\right)x^2 = 27$

3. Ari has eight black socks, w white socks, and six grey socks. Half of his white socks are clean. How many clean white socks does he have if he has 34 socks in all?

   A 10 clean white socks
   B 14 clean white socks
   C 17 clean white socks
   D 20 clean white socks

4. Which value of z makes the equation true?

   $\frac{z}{7} + 13 = 27$

   F $z = 2$
   G $z = \frac{40}{7}$
   H $z = 7$
   I $z = 98$

5. Which of the following is a solution to the equation $\frac{x}{2} + 5 = 33$?

   A $x = 14$
   B $x = 19$
   C $x = 56$
   D $x = 76$

6. While riding in the car along a straight highway, Theresa sees the following sign:

   When Theresa travels half the remaining distance to Phillipsburg, how many miles will she be from Randalville?

   F $58\frac{1}{2}$ miles
   G $87\frac{1}{2}$ miles
   H 88 miles
   I 117 miles

NAEP Objective Test 8.5.4.c continued

7. Which of the following equations does NOT have a solution of $x = 3$?

   A $\left(\frac{2}{3}\right)x = 2$

   B $10x \div 6 = 5$

   C $\left(\frac{1}{3}\right)x + 22 = 23$

   D $3x^2 = 18$

8. Sydney wants to solve the equation $\frac{1}{4}x = 12$. What step should she take first?

   F Subtract $\frac{1}{4}$ from both sides.

   G Subtract 12 from both sides.

   H Multiply both sides by 4.

   I Divide both sides by 4.

9. Which of the following equations has a solution of 5?

   A $\left(\frac{3}{5}\right)x + 7 = 10$

   B $2x - 4 = 8$

   C $\left(\frac{1}{5}\right)x + 17 = 22$

   D $9x \div 3 = 16$

10. Terrence and Wayne are at the train station and see the following:

    > **Traveling light special:**
    > Buy any ticket at full price, take a friend along for only $5.00 more

    Terrence and Wayne decide to buy two tickets and split the cost evenly. The full-price ticket costs $37.00. What will Terrence and Wayne each pay?

    F $18.50

    G $21.00

    H $37.00

    I $42.00

11. Which of the following is the solution to the equation $\frac{t}{3} = 21$?

    A $t = \frac{1}{7}$

    B $t = \frac{7}{3}$

    C $t = 7$

    D $t = 63$

12. What is the value of a in the following equation?

    $$\frac{4}{9} + \frac{2}{3} = a - \frac{1}{2}$$

    F $\frac{7}{14}$

    G $\frac{11}{18}$

    H $\frac{11}{9}$

    I $1\frac{11}{18}$

NAEP Objective Test 8.5.4.e

Select the best answer for questions 1–12. Fill in the correct bubble on your answer sheet.

1. What is the circumference of the circle below? Use 3.14 for π.

   A 15.7 millimeters
   B 25.4 millimeters
   C 31.4 millimeters
   D 62.8 millimeters

2. Carlos puts $150 in a simple interest account. The interest rate is 5% per year. How much interest will Carlos receive after one year?
   F $2.50
   G $3.00
   H $5.00
   I $7.50

3. Sam and his family went cross-country skiing for 3 hours on Thursday. They skied at a rate of 2.7 miles per hour. How far did they ski?
   A 5.7 miles
   B 6 miles
   C 8.1 miles
   D 9 miles

4. A round area rug has a radius of three feet. How much area does the rug cover? Use 3.14 for π.
   F 9.42 square feet
   G 18.85 square feet
   H 28.26 square feet
   I 56.55 square feet

5. Tammy is on a bus trip from her home to her grandmother's home, 150 miles away. Her bus left at 2:00 P.M. By 3:30 P.M., the bus had traveled 90 miles. At that rate, how long will it take to go 150 miles?
   A 1 hour, 30 minutes
   B 1 hour, 50 minutes
   C 2 hours, 30 minutes
   D 2 hours, 50 minutes

6. Fatima earned $6.50 in interest at the end of the year in a simple interest account. The interest rate is 5% per year. How much money did Fatima have in the account at the beginning of the year?
   F $65.00
   G $92.31
   H $122.50
   I $130.00

NAEP Objective Test 8.5.4.e continued

7. Grace earned $14.40 in interest at the end of the year in a simple interest account. The interest rate is 6% per year. How much money did Grace have in the account at the beginning of the year?
   A $86.40
   B $144.00
   C $225.60
   D $240.00

8. What is the circumference of a circle drawn the sidewalk with a diameter of 12 feet? Use 3.14 for π.
   F 18.39 feet
   G 37.68 feet
   H 73.56 feet
   I 113.04 feet

9. An experimental electric car can travel at 60 miles per hour for $7\frac{1}{2}$ hours before it needs to be recharged. How far can the electric car travel before it needs to be recharged?
   A 420 miles
   B 450 miles
   C 730 miles
   D 750 miles

10. Serena puts $400 in a simple interest account. The interest rate is 4.5% per year. How much interest will Serena receive after one year?
    F $18.00
    G $45.00
    H $60.00
    I $88.89

11. What would the area of the circle be if the radius was halved. Use 3.14 for π.

    A 200.96 square meters
    B 100.48 square meters
    C 50.24 square meters
    D 25.12 square meters

12. Hilary spends the day in-line skating at the park. She skates for $3\frac{1}{2}$ hours at an average speed of 12 miles per hour. How many miles does Hilary skate?
    F $32\frac{1}{2}$ miles
    G 35 miles
    H $36\frac{1}{2}$ miles
    I 42 miles

Name _____ Date _____ Class _____

Sample Test A

Select the best answer for questions 1–24. Fill in the correct bubble on your answer sheet.

1. Which two numbers come next in the pattern?

    0, 4, 9, 15, 22, …

    A 23, 24
    B 25, 32
    C 29, 35
    D 30, 39

2. Eric pulls one marble from the box without looking. What is the probability that it will be gray?

    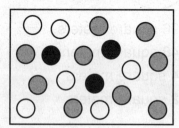

    F $\frac{1}{5}$
    G $\frac{2}{5}$
    H $\frac{1}{4}$
    I $\frac{1}{2}$

3. Identify the best rule for this pattern.

    | 6 | 12 | 18 | 24 | 30 |

    A The numbers are increasing by 2.
    B The numbers are increasing by 6.
    C The numbers are doubling.
    D The numbers are halving.

4. Which property is demonstrated by the equation
   8 + (9 + 3) = (8 + 9) + 3?

    F Commutative Property of Addition
    G Associative Property of Addition
    H Distributive Property
    I Property of Opposites

5. Darcy recorded the temperature for the past five days. Order these temperatures from least to greatest.

    | Day 1 | 75.2° |
    | Day 2 | 74.02° |
    | Day 3 | 75.36° |
    | Day 4 | 76.05° |
    | Day 5 | 74.5° |

    A 74.02°, 74.5°, 75.2°, 75.36°, 76.05°
    B 74.02°, 74.5°, 75.36°, 75.2°, 76.05°
    C 74.02°, 76.05°, 75.2°, 74.5°, 75.36°
    D 76.05°, 75.36°, 75.2°, 74.5°, 74.02°

6. Gilda completed 60% of her assignment. Which shows another way to describe this number?

    F 0.06
    G $\frac{60}{10}$
    H $\frac{3}{5}$
    I 6.0

Holt Mathematics Grade 6

Sample Test A continued

7. One International Nautical Mile is equal to 1.852×10^3 meters. Which of the following shows the number of meters in standard form?

A 0.001852
B 185.2
C 1,852
D 1,852,000

8. What would be the area of the circle if the radius were halved? Use 3.14 for π.

F 200.96 square meters
G 100.48 square meters
H 50.24 square meters
I 25.12 square meters

9. Which figure has exactly 5 vertices?

A hexagon
B pentagon
C decagon
D polygon

10. Identify the figure shown.

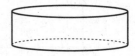

F square pyramid
G cube
H cylinder
I cone

11. On a sunny day, a saguaro cactus casts a shadow that is 56 feet long. At the same time a rabbit casts a shadow that is 4 feet long. What is the height of the cactus?

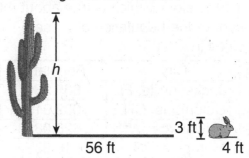

A 12 feet
B 42 feet
C 48 feet
D 60 feet

12. Ron has 36 markers, 48 erasers, and 18 pens. He will put them in bags with the same number of each item. What is the greatest number of bags he can make?

F 2
G 6
H 12
I 16

Name _____ Date _____ Class _____

Sample Test A

13. What are the coordinates of point Q?

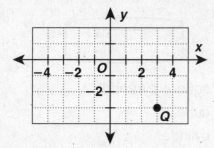

A (2, 2)
B (2, −3)
C (3, −3)
D (−3, 3)

14. The 1990 population of 5 cities are listed below. Which is true about the sum of the population of these 5 cities in 1990?

City	Population
Jacksonville, FL	635,230
Columbus, OH	632,910
Milwaukee, WI	628,088
Memphis, TN	610,337
Boston, MA	574,283

F more than 3,150,000
G about 5,300,000
H between 3,000,000 and 3,100,000
I less than 3,000,000

15. Bill rides his bicycle $1\frac{2}{3}$ miles every day. Julia rides $1\frac{1}{4}$ times as far as Bill does. How many miles does Julia ride her bicycle every day?

A $\frac{5}{12}$ mile
B $2\frac{1}{12}$ miles
C $2\frac{3}{7}$ miles
D $2\frac{5}{12}$ miles

16. What is the sample space for an experiment of tossing two coins?

F H, T
G HH, TT
H HH, TT, HT
I HH, TT, HT, TH

17. Hannah wants to measure the height of a dime. Which unit should she use to get the the most precise measurement?

A millimeter
B inch
C eighth-inch
D centimeter

18. What is the missing angle measure of the quadrilateral shown in the figure?

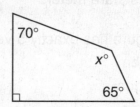

F 135°
G 125°
H 115°
I 105°

Sample Test A continued

19. Which lines are parallel?

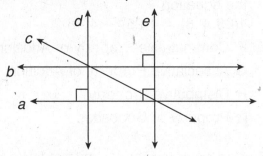

   A *a* and *b*
   B *c* and *d*
   C *a* and *e*
   D *b* and *c*

20. Which shows a 45° clock-wise rotation of the figure shown below?

   F   H

   G   I

21. Rhonda wants to buy 60 pencils for her class. She priced the pencils at four stores. Which is the best buy?

   A Store A: 15 pencils for $2.62
   B Store B: 10 pencils for $1.80
   C Store C: 6 pencils for $0.90
   D Store D: 5 pencils for $0.80

22. What is the solution of the equation $2x + 3 = 4x - 7$?

   F $x = 5$
   G $x = 4$
   H $x = 3$
   I $x = 2$

23. LaToya bought a new pair of running shoes that cost $49.95. The sales tax rate is 6%. To the nearest cent, how much did she spend, including tax?

   A $79.92
   B $55.95
   C $52.95
   D $50.55

24. Bags of chips sell for $0.30 each. Which table shows the cost (*c*) of 5, 10, 15, and 20 bags (*b*) of chips?

   $c = 0.30b$

F
b	1	2	3	4
c	$0.30	$0.60	$0.90	$1.20

G
b	5	10	15	20
c	$0.35	$0.40	$0.45	$0.50

H
c	$5	$10	$15	$20
b	6	3	2	1.5

I
b	5	10	15	20
c	$1.50	$3.00	$4.50	$6.00

Name _____ Date _____ Class _____

Sample Test B

Select the best answer for questions 1–24. Fill in the correct bubble on your answer sheet.

1. Which is the missing number in the sequence?

    54, 50, 46, ____, 38, 34

    A 44
    B 42
    C 39
    D 20

2. Rasheed pulls one marble from the box without looking. What is the probability that it will NOT be white?

    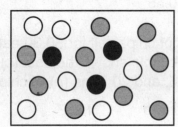

    F $\frac{1}{6}$
    G $\frac{1}{3}$
    H $\frac{1}{2}$
    I $\frac{2}{3}$

3. Identify the best rule for this pattern.

    The numbers are increasing by 7.

    A 1, 7, 49, 343
    B 4; 47; 477; 4,777
    C 356, 349, 342, 335
    D 0, 7, 14, 21

4. Which property is demonstrated by the equation
    4(25 + 8) = 4(25) + 4(48)?
    F Commutative Property of Addition
    G Associative Property of Addition
    H Distributive Property
    I Property of Opposites

5. Meg recorded the weights of five half dollars. Order the weights from least to greatest.

Year Made	Weight (g)
1992	11.08
1993	10.76
1994	11.34
1995	10.9
1996	11

    A 10.76, 10.9, 11, 11.34, 11.08
    B 10.9, 10.76, 11.08, 11.34, 11
    C 10.76, 10.9, 11, 11.08, 11.34
    D 11, 10.9, 10.76, 11.08, 11.34

6. Five-eighths of Patrick's baseball team bats at least twice every game. What percent of the team bats at least twice every game?
    F 62.5%
    G 625%
    H 62%
    I 63%

Copyright © by Holt, Rinehart and Winston.
All rights reserved.

Holt Mathematics Grade 6

Sample Test B continued

7. The Nile River runs 4,160 miles. Which of the following shows the number of miles written in scientific notation?

   A 0.416×10^4
   B 41.6×10^3
   C 4.16×10^2
   D 4.16×10^3

8. The sides of a square are 2 centimeters long. What would be the area of the square if the side lengths were tripled?

   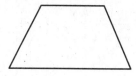
   2 cm

   F 36 centimeters
   G 36 square centimeters
   H 64 centimeters
   I 64 square centimeters

9. Which type of triangle has exactly two equal sides?

   A isosceles
   B equilateral
   C obtuse
   D scalene

10. Identify the figure shown.

    F pentagon
    G trapezoid
    H hexagon
    I octagon

11. What is the height of the tree?

    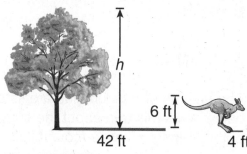

    A 28 feet
    B 60 feet
    C 63 feet
    D 126 feet

12. The Drama Club has 15 girls and 27 boys. What is the largest number of skit groups possible if each group must have the same number of girls and the same number of boys and each student must perform a skit?

    F 3 groups
    G 6 groups
    H 9 groups
    I 45 groups

Sample Test B

13. What are the coordinates of point *P*?

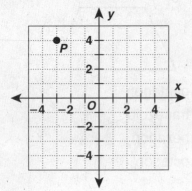

- A (4, −3)
- B (−3, 4)
- C (−4, 3)
- D (3, −4)

14. The heights of 5 mountains are listed below. Which is true about the sum of the heights of these mountains?

Mountain	Height (ft)
Dhaulagiri	26,810
Nanga Parbat	26,660
Annapurna	26,504
Gasherbrum	26,470
Xixabangma	26,286

- F more than 150,000
- G about 1,030,000
- H between 100,000 and 120,000
- I between 130,000 and 135,000

15. Objects on Neptune weigh $1\frac{1}{8}$ times as much as they weigh on Earth. How much would a $9\frac{1}{3}$-lb backpack weigh on Neptune?

- A $10\frac{1}{2}$ lb
- B 10 lb
- C $9\frac{1}{24}$ lb
- D $8\frac{8}{24}$ lb

16. What is the sample space for an experiment of tossing a coin and then spinning a spinner divided into three equal sections labeled A, B, and C?

- F H, T, A, B, C
- G HA, TB, HC
- H HA, TA, HB, TB, HC, TC
- I HH, TT, HT, TH, AB, BC, AC, AA, BB, CC

17. Four students measured the height of the classroom door. Which of these measurements is the most precise?

- A $6\frac{1}{2}$ feet
- B 6 feet, 7 inches
- C 2 yards, $\frac{1}{2}$ foot
- D 2 yards, $6\frac{3}{4}$ inches

18. What is the value of *x* in the parallelogram shown?

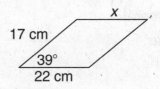

- F 141 centimeters
- G 39 centimeters
- H 17 centimeters
- I 22 centimeters

Sample Test B continued

19. Which lines are parallel?

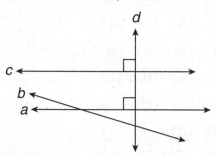

- **A** a and b
- **B** b and d
- **C** a and c
- **D** c and d

20. Which shows a reflection of the figure shown below?

F ⊠ H ⊠

G ⊠ I ◇

21. Roberta wants to buy 64 ounces of sour cream. She priced the sour cream at four stores. Which is the best buy?

- **A** Store A: 16 ounces for $1.50
- **B** Store B: 32 ounces for $2.74
- **C** Store C: 8 ounces for $0.55
- **D** Store D: 4 ounces for $0.43

22. What is the solution of the equation $4x - 2 = x + 7$?

- **F** $x = 5$
- **G** $x = 4$
- **H** $x = 3$
- **I** $x = 2$

23. Kendra paid $48 for a sweater. If she bought it on sale at a 40% discount, what was the original price??

- **A** $19.92
- **B** $28.80
- **C** $67.20
- **D** $80.00

24. The speed limit near Bev's house is 55 miles per hour. Which table shows the number of miles (m) driven in 1, 2, 3, and 4 hours (h) if a constant speed of 55 miles per hour is maintained?

$m = 55h$

F
h	1	2	3	4
m	55	110	165	220

G
m	1	2	3	4
h	25	50	100	150

H
h	1	20	3	4
m	55	60	65	70

I
h	1	2	3	4
m	55	85	115	145

Name _____ Date _____ Class _____

Sample Test Answer Sheet

Sample Test A

1. Ⓐ Ⓑ Ⓒ Ⓓ
2. Ⓕ Ⓖ Ⓗ Ⓘ
3. Ⓐ Ⓑ Ⓒ Ⓓ
4. Ⓕ Ⓖ Ⓗ Ⓘ
5. Ⓐ Ⓑ Ⓒ Ⓓ
6. Ⓕ Ⓖ Ⓗ Ⓘ
7. Ⓐ Ⓑ Ⓒ Ⓓ
8. Ⓕ Ⓖ Ⓗ Ⓘ
9. Ⓐ Ⓑ Ⓒ Ⓓ
10. Ⓕ Ⓖ Ⓗ Ⓘ
11. Ⓐ Ⓑ Ⓒ Ⓓ
12. Ⓕ Ⓖ Ⓗ Ⓘ
13. Ⓐ Ⓑ Ⓒ Ⓓ
14. Ⓕ Ⓖ Ⓗ Ⓘ
15. Ⓐ Ⓑ Ⓒ Ⓓ
16. Ⓕ Ⓖ Ⓗ Ⓘ
17. Ⓐ Ⓑ Ⓒ Ⓓ
18. Ⓕ Ⓖ Ⓗ Ⓘ
19. Ⓐ Ⓑ Ⓒ Ⓓ
20. Ⓕ Ⓖ Ⓗ Ⓘ
21. Ⓐ Ⓑ Ⓒ Ⓓ
22. Ⓕ Ⓖ Ⓗ Ⓘ
23. Ⓐ Ⓑ Ⓒ Ⓓ
24. Ⓕ Ⓖ Ⓗ Ⓘ

Sample Test B

1. Ⓐ Ⓑ Ⓒ Ⓓ
2. Ⓕ Ⓖ Ⓗ Ⓘ
3. Ⓐ Ⓑ Ⓒ Ⓓ
4. Ⓕ Ⓖ Ⓗ Ⓘ
5. Ⓐ Ⓑ Ⓒ Ⓓ
6. Ⓕ Ⓖ Ⓗ Ⓘ
7. Ⓐ Ⓑ Ⓒ Ⓓ
8. Ⓕ Ⓖ Ⓗ Ⓘ
9. Ⓐ Ⓑ Ⓒ Ⓓ
10. Ⓕ Ⓖ Ⓗ Ⓘ
11. Ⓐ Ⓑ Ⓒ Ⓓ
12. Ⓕ Ⓖ Ⓗ Ⓘ
13. Ⓐ Ⓑ Ⓒ Ⓓ
14. Ⓕ Ⓖ Ⓗ Ⓘ
15. Ⓐ Ⓑ Ⓒ Ⓓ
16. Ⓕ Ⓖ Ⓗ Ⓘ
17. Ⓐ Ⓑ Ⓒ Ⓓ
18. Ⓕ Ⓖ Ⓗ Ⓘ
19. Ⓐ Ⓑ Ⓒ Ⓓ
20. Ⓕ Ⓖ Ⓗ Ⓘ
21. Ⓐ Ⓑ Ⓒ Ⓓ
22. Ⓕ Ⓖ Ⓗ Ⓘ
23. Ⓐ Ⓑ Ⓒ Ⓓ
24. Ⓕ Ⓖ Ⓗ Ⓘ

Copyright © by Holt, Rinehart and Winston.
All rights reserved.

Holt Mathematics Grade 6